BOOKS BY JOAN ELMA RAHN

Grocery Store Botany
How Plants Are Pollinated
How Plants Travel
Seeing What Plants Do
More About What Plants Do
Alfalfa, Beans & Clover
Grocery Store Zoology
The Metric System
Seven Ways to Collect Plants
Traps and Lures in the Living World

Traps & Lures in the Living World

Joan Elma Rahn

Traps & Lures in the Living World

ATHENEUM 1980 NEW YORK

LIBRARY OF CONGRESS CATALOGING IN PUBLICATION DATA

Rahn, Joan Elma, date
Traps and lures in the living world.

SUMMARY: Introduces the many ways plants and animals capture animals for food, pollination, mating, and other purposes.

1. Predation (Biology)—Juvenile literature.
2. Animals, Food habits of—Juvenile literature.
3. Insectivorous plants—Juvenile literature.
4. Fungi, Predatory—Juvenile literature.
[1. Predatory animals. 2. Animals—Food habits.
3. Insectivorous plants] I. Title.
QL758.R33 574.5 79-22699
ISBN-0-689-30766-7

Published simultaneously in Canada by
McClelland & Stewart, Ltd.
Manufactured by American Book–Stratford Press, Inc.
Saddlebrook, New Jersey
Designed by M. M. Ahern
First Edition

Contents

"Will you walk into my parlor?"
said the spider to the fly;
" 'Tis the prettiest little parlor
that ever you did spy."

The Spider and the Fly
Mary Howitt
1799–1888

INTRODUCTION

When an animal needs something—food, for example—it usually goes out and gets it, but a few animals wait for food to come to them. Some of these animals build traps that greatly increase the chances that once another animal has come into the area it will be prevented from leaving and so will be easier to catch.

Some of these waiting animals lure their prey to them. The lure may be something that the prey animal needs or wants, but usually it is a "fake"—like a fisherman's lure that is made of feathers but looks enough like a real fish that a larger fish will bite it. Depending on the species involved, the lure may mimic food, an appropriate mate, or even a place to rest.

Some animals catch their prey with traps alone, some with lures alone, and some with a combination of traps and lures.

In the matter of food, most green plants are self-sufficient as long as they have plenty of sunshine, carbon dioxide, water, and minerals. A few green plants, however, do much better if they have an occasional animal that they can digest. These carnivorous plants catch small animals in traps that are modified leaves.

When it comes to cross-pollination, plants, which do not have the power to move about as animals do, are dependent on something other than themselves for the transport of pollen from flower to flower. If the pollinator is an animal, the flower nearly always has a lure that attracts the animal. A few flowers even trap their pollinators for a while.

This book presents a short selection of traps and lures in the plant and animal kingdoms.

1

"Will You Walk into My Parlor?"

Spiders eat animals—mostly insects, but other small animals as well. Many spiders trap these animals by building webs in which the victims become snared.

SPIDER SILK

The threads of a spider web are made of silk produced by the spider itself. Spider silk is quite strong. Small animals caught in it are usually unable to break the threads. Spider silk is also very fine, and unless the light shines on it from just the right direction, a spider's web may be difficult to see. Spider silk is even finer than silkworm silk, which is what we use in the manufacturing of silk clothing.

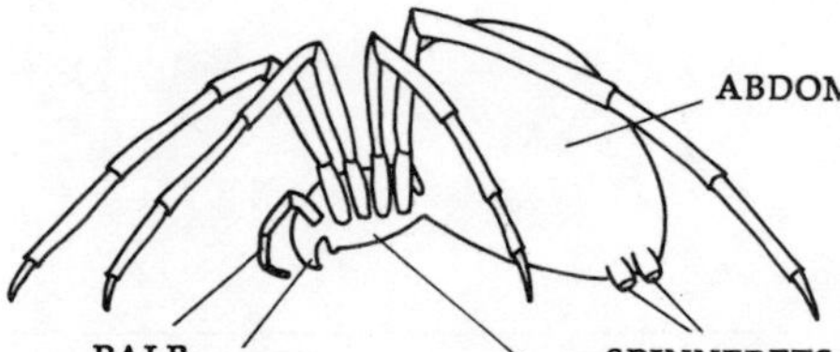

The body of a spider is divided into two main parts: cephalothorax and abdomen. Four or six spinnerets, from which silk threads are spun, are near the rear tip of the abdomen. Spiders have four pairs of legs. They also have a pair of palps, which are modified mouth parts, but which often look like short legs. The fangs eject poison into animals trapped in the web. Only the appendages of the left side of the body are shown here.

Spiders produce several kinds of silk in silk glands located in the abdomen. Each spider has several types of silk glands, each producing a different type of silk. The silk produced in the silk glands is a liquid. This liquid can move from the glands into four or six *spinnerets* near the rear of the spider's abdomen. The spider controls which glands are open and which glands are closed. When a gland is open, the liquid silk from that gland comes out of the spinneret. As the liquid silk becomes stretched, it solidifies into a silken thread.

Different kinds of spiders make different kinds of webs. We will consider first the *orb web,* then some others. The orb web is made by the spiders called *orb weavers.* Their web is the kind you usually see illustrating Halloween or ghost stores or in pictures of dusty attics or basements. It is a flat, two-dimensional web with a spiral thread and with several lines radiating from the center. These lines are called *radii* (singular, *radius*). Most orb webs stand nearly vertical with just a little tilt.

The following kinds of silk are used in the building of an orb web: dragline silk, attachment disks, elastic silk, viscid silk, and hackled bands.

Dragline silk is a dry silk. As a spider walks or jumps from place to place it continuously spins a *dragline.* Occasionally the spider attaches the dragline to some solid object. Thus secured, the dragline functions as a lifeline. If the spider should jump or

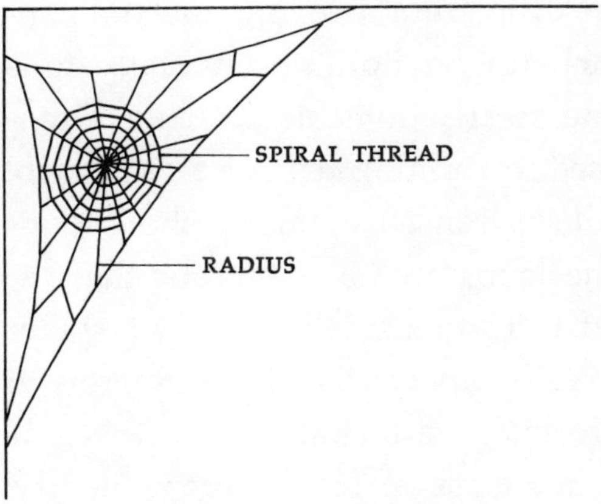

The general form of an orb web.

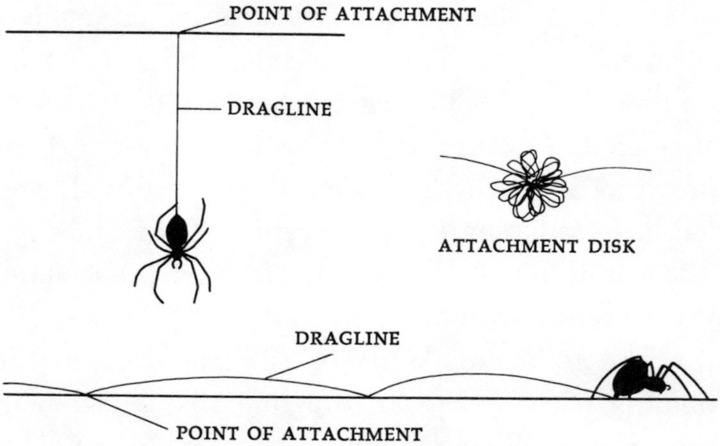

Wherever it goes, a spider spins out dragline silk behind it and attaches the line at intervals. Because the silk is so fine, we ordinarily do not see it unless we look carefully for it.

fall from a great height, it can stop its fall by stopping the spinning of the dragline. Then the spider can climb up the dragline to safety. Perhaps you have seen a spider drop from a ceiling and then appear to climb back up on nothing. Actually it is climbing on its fine, nearly invisible dragline. Dragline silk is also used in forming part of an orb web.

Attachment disks consist of many short loops of silk thread. The loops are so numerous that together they appear to form a small disk. As a spider walks about, it occasionally touches its spinnerets to some object and forms an attachment disk there. If you keep a spider in a glass jar for a short while, the spider will spin some draglines and secure them to

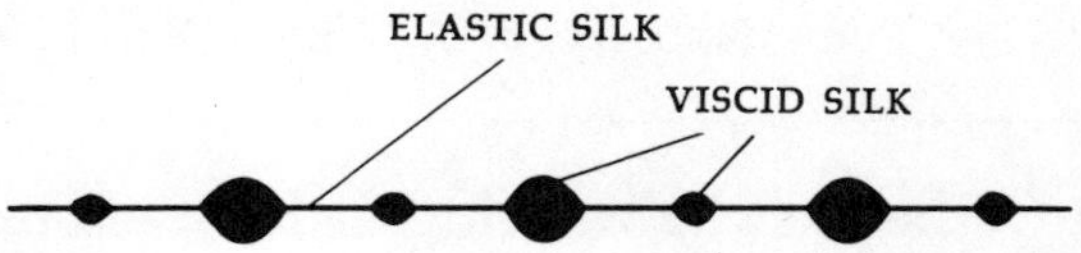

A short length of the spiral thread much enlarged.

the sides of the jar by attachment disks. To the unaided eye, the disks appear as small white dots.

Elastic silk is silk that stretches. It forms the spiral part of the completed orb web. *Viscid* (pronounced VISS-id) *silk* is sticky silk; it occurs as drops, evenly spaced, on the elastic silk. The elastic and viscid silk form the trapping portion of the orb web. The parts of the web made from dragline silk merely support the elastic threads with their viscid drops.

Hackled bands are flat threads with which some orb weavers finish off their webs.

Although in some species both male and female spiders spin webs, a spider web that you find is more likely to have been made by a female spider. Usually male spiders make webs only when they are young (if at all) and not during their adult lives. For this reason, a spider spinning a web will be referred to as *she* and the trapped victim—which is likely to be of either sex—as *it*.

THE ORB WEB

One of the common orb weavers in gardens in the United States and Canada, is the *black and yellow argiope.*

The argiope begins the weaving of her web by selecting a spot several feet above the ground. Often it will be a small open space over a path or a narrow stream. These are places where insects fly and thus are likely to become snared in the web.

The spider faces into the wind, raises her abdomen, and spins some dragline silk. The free end of the line floats on the breeze and may become attached to something. If the end does not become attached to an object, the spider hauls the line back to her and eats it. But if the free end touches another object and becomes attached to it, the spider tightens the line and attaches the other end at her location. This line, the first line of the orb web, is called the *bridge line*. The spider may strengthen the bridge line by crossing it once or twice and laying down more threads as she goes.

The spider then moves to a spot in the middle of the bridge, attaches a new line to one of the threads of the bridge, and drops downward, spinning as she falls. This stretches and pulls down the thread of the bridge, which, together with the newest thread, forms a Y. The center of the Y will be the center, or *hub*, of the web. The threads of the Y itself form the first three radii of the web.

Where she lands, the spider attaches the thread, then returns to the hub. Here she attaches a new thread that will be the next radius of the web. As she spins out this thread, she walks along one of the radii, then along whatever is supporting the web. All

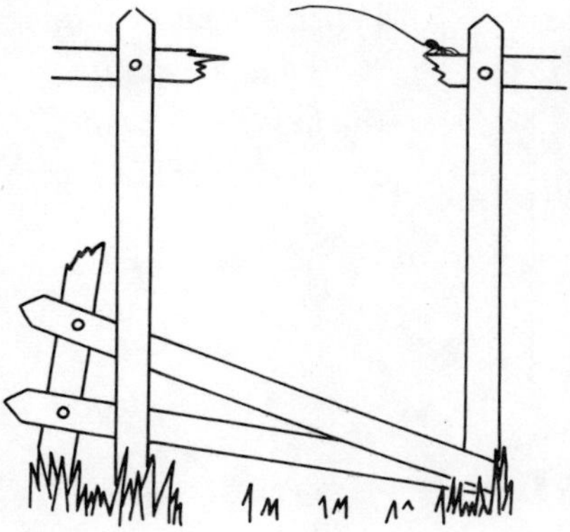

A spider begins building a web by letting a dragline float on the breeze.

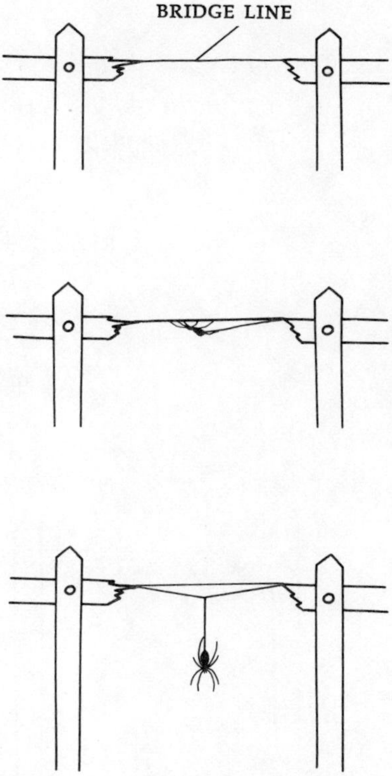

Above: Bridge line completed. Middle: Spider strengthens the bridge line by adding a second line to it. Below: Spider drops a vertical line.

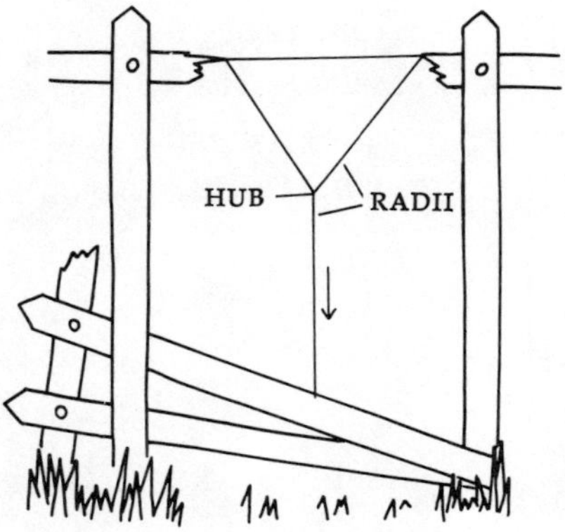

When the vertical line is attached below, the first three radii of the web are in place.

The spider lays down the fourth radius of the web.

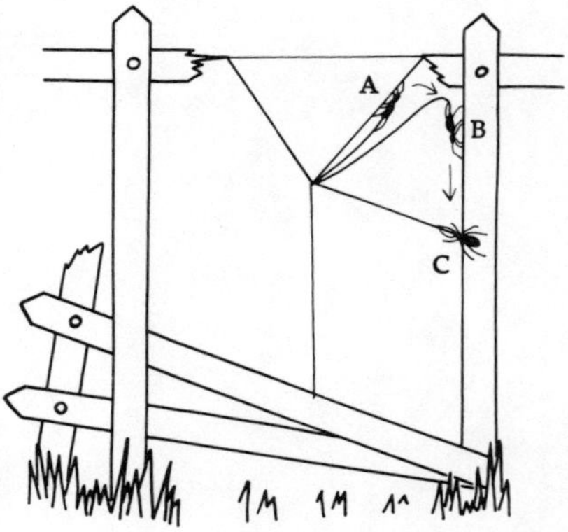

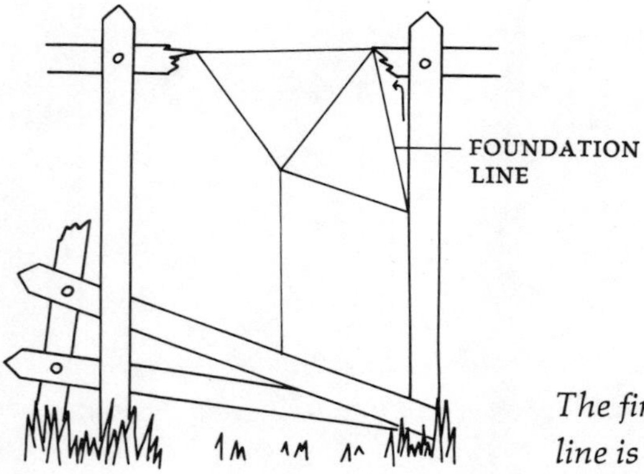

The first foundation line is formed.

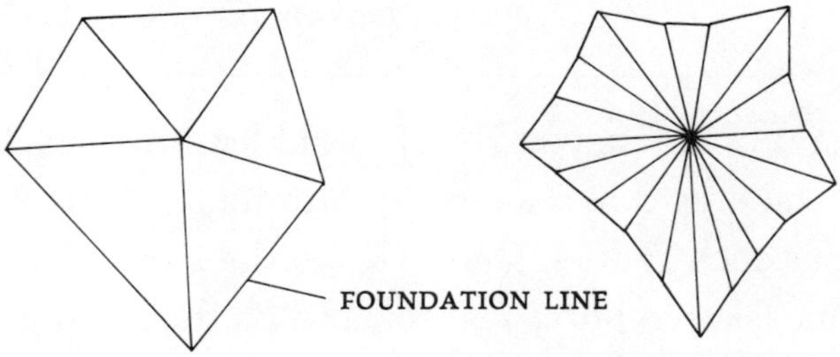

After completing the foundation lines and several radii (*left*), *the spider lays down additional radii among them* (*right*).

the time she keeps the new line free from entanglements by guiding it with one of her hind legs. Selecting a spot, she tightens the line and attaches it. Then, still spinning, she returns to the bridge and connects the line to it. This is the first of several *foundation lines.*

The spider returns to the hub several more times and forms in the same way a few more radii and additional foundation lines. The bridge line and the foundation lines outline the web. The spider then forms more radii among the old ones. When all the radii are in place, the spaces between them are nearly equal, but not exactly so. The radii usually are a little closer together in the bottom half of the web than in the top half.

At the center of the web, the spider strengthens the web by adding threads that connect the radii. This forms a *notched zone.* The radii may become quite irregular and difficult to follow through the notched zone. The spider leaves a *free zone* around the notched zone. In the free zone, no new threads are spun, but beyond it, the spider attaches a line that she lays down as one continuous spiral. As she crosses each radius, she attaches the spiral thread to it. This is a *temporary spiral,* and its turns are farther apart than will be those of the permanent spiral that will be formed next. Some spiders proceed clockwise when making the temporary spiral, others counterclockwise.

Up to this point, the spider has laid down only dry silk. Now she returns to the free zone, following the temporary spiral as she does. She removes the dry threads of the temporary spiral, rolling up the thread and either eating it or discarding it. At the same time, she spins a new spiral of elastic and viscid silk. Both the elastic silk and the viscid silk come out

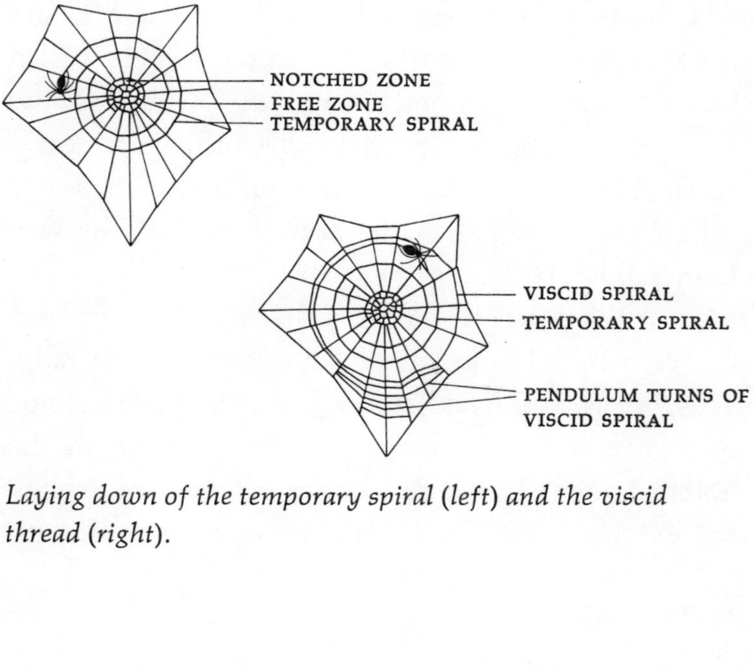

Laying down of the temporary spiral (*left*) *and the viscid thread* (*right*).

of the spinnerets as continuous threads. As she attaches the thread to each radius, she pulls it tight and then releases it with a little snap. This breaks the viscid silk into tiny droplets that adhere to the elastic silk. In some webs the drops of viscid silk are just barely large enough to be seen with the naked eye.

This spiral thread of elastic and viscid silk is called the *viscid spiral* or the *catching thread.* It is also called the *permanent spiral,* but it is not truly permanent for it may be severely damaged by the struggles of an insect caught in it. Even if it is not torn this way, the spider usually removes it when it grows old and replaces it with a new one.

The turns of the viscid spiral are more numerous and are closer together than were those of the temporary spiral. The spider generally forms the viscid spiral by proceeding in a direction opposite to that in which she formed the temporary spiral. That is, she proceeds in a counterclockwise direction if she had formed the temporary spiral in a clockwise direction. However, from time to time she may reverse her direction and make *pendulum turns.* Pendulum turns are quite common in the bottom part of the web.

The argiope spider then adds a broad zigzag ribbon made from hackled bands. This ribbon was called a *stabilimentum* because some persons believed that it stabilized the web, but actually it does not. Because it is conspicuous, it probably draws the attention of insects away from the spider when she

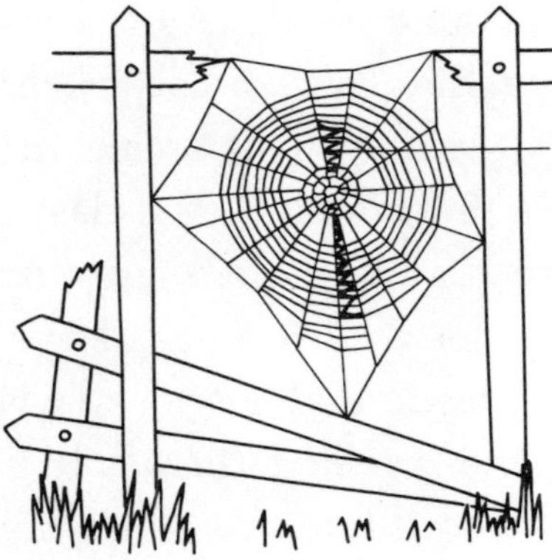

Completed orb web.

sits waiting at the center of the web. It may also make the spider less conspicuous to birds or other animals that eat spiders.

Not all spiders wait on the web itself. Some hide under nearby foliage, which they use as a *retreat*. They spin a line, called a *signal line,* from the web to their retreats. A change in the tension of the signal line usually indicates that an insect has become caught in the web.

When an insect flies into the web, it becomes stuck on the drops of viscid silk. Touching just a few drops may not be enough to hold medium-sized or large insects, but as the insect struggles to free itself, it is likely to touch more and more drops of viscid silk. Furthermore, the thrashing of the insect causes the elastic threads to stretch. This brings the insect against even more turns of the viscid spiral, and it is likely to become so entangled that it cannot escape. Because spider silk is strong, few insects are able to break it. However, the wings of moths and butterflies are covered with loosely attached scales. If only the wings of these insects touch the viscid silk, the scales break off and the insect is free to fly away.

The struggles of a trapped insect cause the web to vibrate, and this alerts the spider, who usually comes immediately to the prey. An argiope spider then quickly wraps the prey in a special silk called *swathing silk*. She touches her spinnerets to the insect, and, as she spins out the swathing silk, she holds the insect with one pair of her legs, and with other legs she turns it over and over. With each turn, the insect is covered with another layer of silk. Then she injects a poison into the prey by biting it. The poison paralyzes the insect but does not kill it. Because the insect is alive, it will remain fresh for some time, and the spider can return to it whenever she becomes hungry.

Some spiders paralyze their prey first, then wrap it. The spider poison takes a very short time to paralyze the prey, but if the prey is a bee that stings or an insect that squirts out a poisonous liquid, there could be time enough for the prey to harm the spider. Because argiopes wrap their prey so gently, however, the prey is not annoyed until it is securely wrapped and helpless; then the argiope can bite it with no danger to herself.

If the damage done to the web by a thrashing insect is small, the spider may repair it. If the dam-

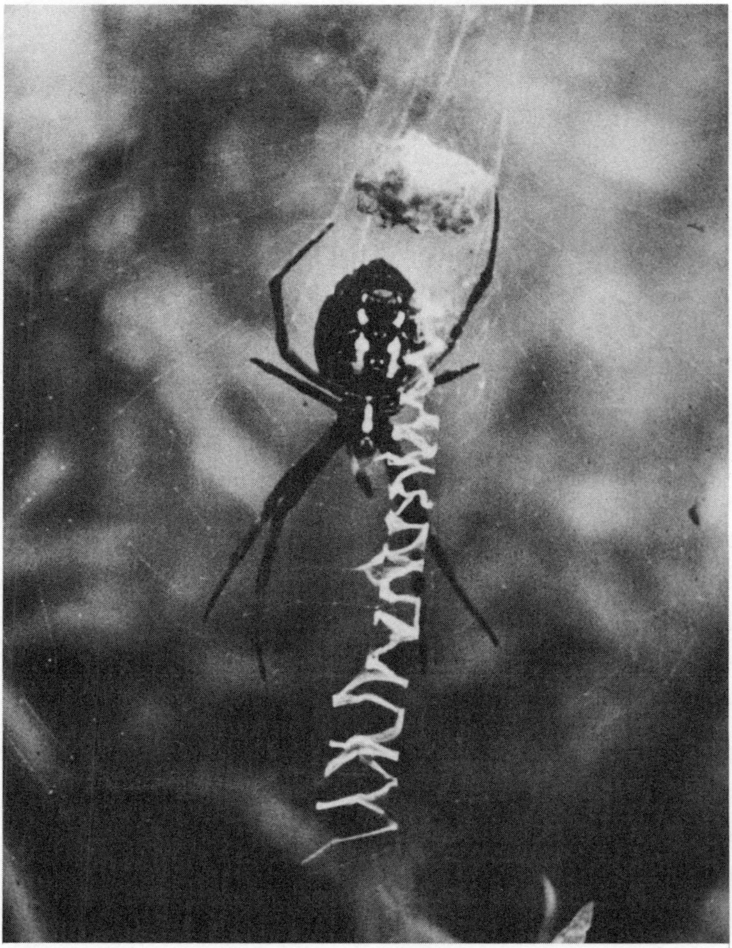

The black and yellow argiope on her web. Above her is one of her victims wrapped in swathing silk. The stabilimentum, which appears as a zigzag line, is made of hackled bands. The viscid spiral and the radii are barely visible.

age is great, she probably will spin a new web. Some spiders build a new web every night and destroy it in the morning whether something was caught in it or not.

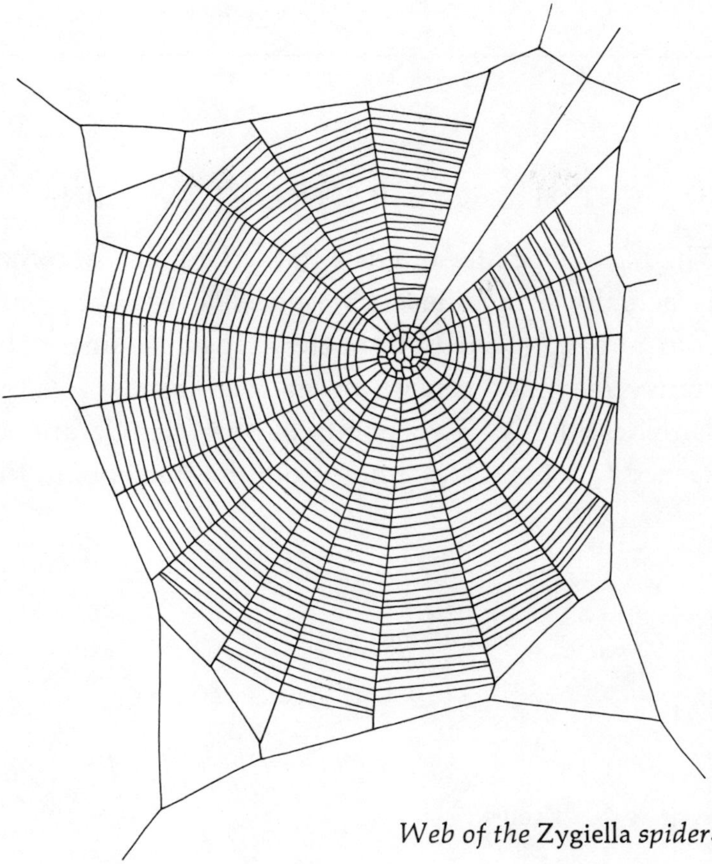

Web of the Zygiella *spider.*

Orb weavers other than argiopes produce webs that are similar to the argiope's web but with some variations. The orb web of the *Zygiella* spider, for instance, has no true spiral thread. Instead, the catching thread is all laid down in pendulum fashion, and two adjacent sectors—on either side of one radius—are left free. The radius in the center of this *free sector* is the spider's signal line. The *Zygiella* spider hides in foliage near the end of the signal line.

OTHER SPIDER WEBS

The *triangle spider* builds a flat, triangular web that is actually a modified orb web. The web has only four radii connected by shorter transverse lines. The transverse lines consist of hackled bands covered with viscid silk. Two points of the triangle are attached to a twig. Extending from the third point, the

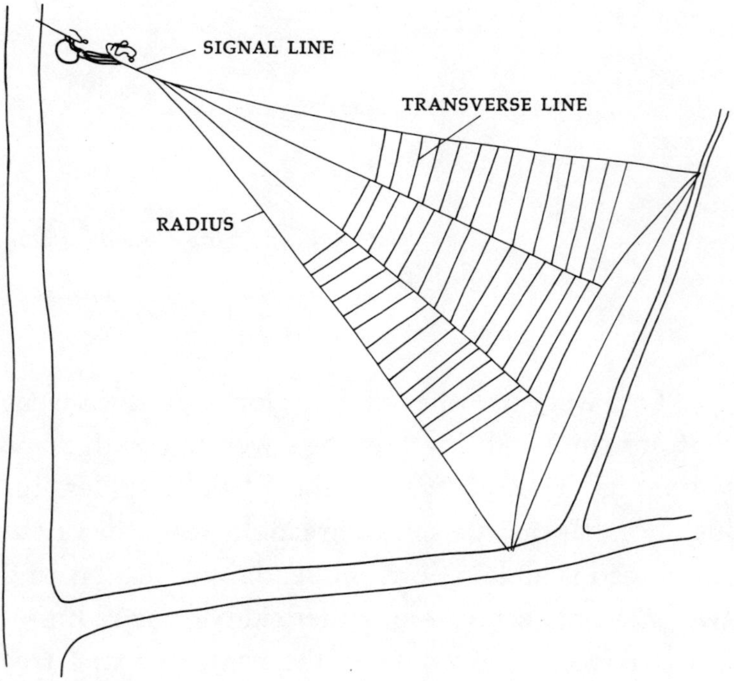

Web of the triangle spider. The spider waits in a break in the signal line.

one at which the four radii meet, is a signal line, attached to another twig. The triangle spider waits on this line—or perhaps it would be better to say that she *is* part of the line, for she cuts it and holds both ends. She has actually become part of her own signal line, which she pulls tight. Then she waits.

When an insect flies or falls into the web, the triangle spider lets the line go slack for an instant. This causes the web to spring forward, and the prey becomes entangled in it. The spider tightens the line quickly. She may repeat these moves once or twice. Then, with the prey firmly entangled, the spider moves down the web and wraps the victim in swathing silk.

Stick spiders live in tropical regions. Their webs are about as simple as a spider web could be. The stick spider places a single line across an open space between two trees and then covers the central part of the line with viscid silk. She assumes a position near one end of the line and holds it tight. So thick is the viscid silk on the line that it is readily visible to small insects, which apparently mistake it for a slender twig on which they might rest safely. Instead, when the insect lands, it becomes caught on the viscid silk. The stick spider then operates her line much as the triangle spider operates her triangle web. She lets the line go slack, and the insect becomes more firmly held by the loose line. The spider then moves along the line and takes her prey.

Grass spiders produce webs that can be seen on

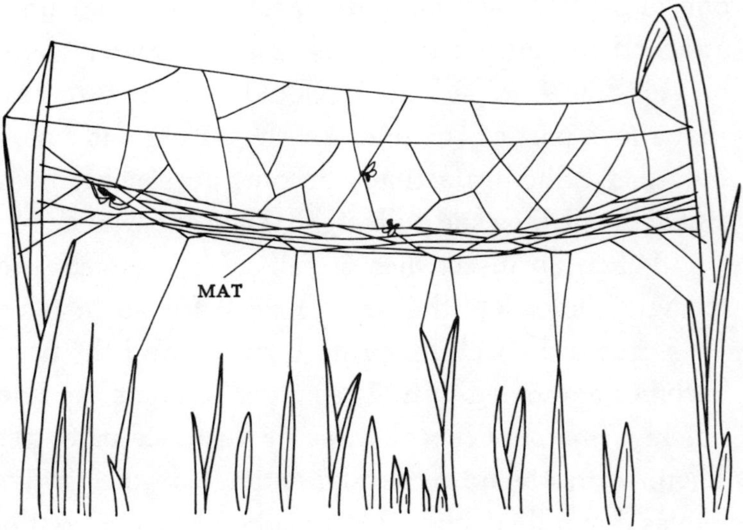

The web of a grass spider. This drawing is highly diagrammatic. The mat of the web is made of many closely woven threads that form a silken sheet.

lawns, especially early on late summer or autumn mornings. If dew covers them, they sparkle in the sunlight. The web of a grass spider consists of a mat of threads of dry silk attached to blades of grass or other short plants. Above the mat, which forms the main part of the web, are a few irregularly placed lines that form a nearly invisible network. Insects flying into this network fall onto the mat. This alerts the spider, who rushes out to capture the victim. Because the web contains no elastic or viscid silk, the spider must be quick, for many insects can free themselves if the spider does not get to them soon.

Some grass spiders build their own retreat—a

silken funnel that extends from the mat down among the grass blades. Others wait on the mat itself.

Steatoda borealis is a spider that lives in tree trunks and in buildings. It builds a horizontal web that is a loosely woven sheet suspended by long lines from a fairly straight support. Additional lines anchor it below. All these lines are dry silk. Hanging from the sheet are a few lines of elastic silk stretched tight and attached below. The lower portion of each elastic thread is covered with drops of viscid silk. This web traps, primarily, walking insects rather than flying ones. When an insect blunders into the viscid drops, its attempts to escape break the elastic threads, which, like rubber bands, become shorter when no longer stretched. The shortening lines pull the insect above the ground and hold it there until the spider hauls it up. A struggling insect may well

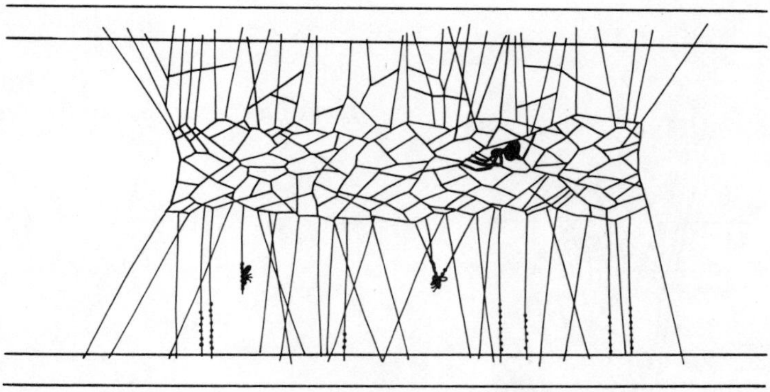

Web of Steatoda borealis. *One insect has broken a single elastic thread. Another has broken three and become entangled in all of them.*

break several elastic threads and so become even more firmly held.

Common *house spiders* build irregular webs that have no particular pattern. These *cobwebs* fill corners of rooms, spaces under beds or behind books in bookcases, and many other places that are left undisturbed by housekeepers for at least a few days. These cobwebs are made of only dry silk. Like the grass spiders, house spiders must be quick about capturing an insect or it will free itself.

THE CADDIS-FLY NET

Very few animals other than spiders spin silken traps. However, caddis-fly larvae, which live in water, spin a funnel-shaped net attached to twigs or other debris in a stream. Both the wide, open mouth

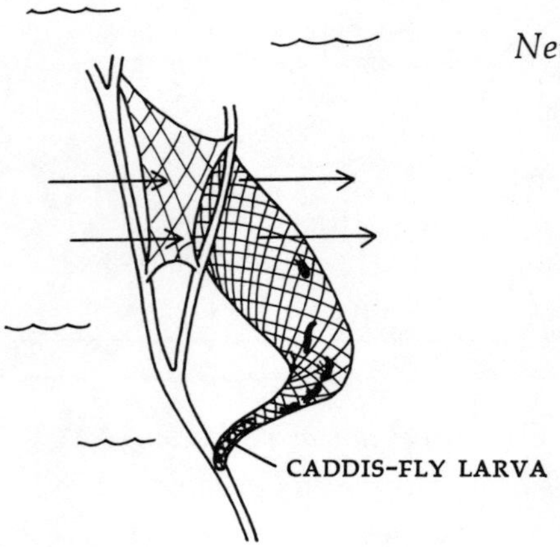

Net of a caddis-fly larva.

of the funnel and the narrow, closed end are pointed upstream. Water flows into the open mouth and then out through the fine meshes of the silk. This strains out small organisms, which fall down to the lower part of the net, where the larva lives. The caddis-fly larva's home thus is also a trap that catches and brings its food to it.

THE ANT LION'S PITFALL

Not all traps are woven of silk. In its larval stage, the ant lion constructs a type of trap called a *pitfall.* A pitfall is a trap into which a victim easily falls, but from which it is difficult to escape. Human beings usually construct a pitfall by digging a deep hole in firm soil; the sides are vertical, and a victim falling into the pit usually cannot climb up such steep sides.

An ant lion, however, constructs its pitfall in fine, dry soil or dry sand. If you have ever dug holes in the dry portion of a sandy beach, you know that it is not possible to dig a pit there with vertical sides. You can easily dig a hole the sides of which have a very shallow slant. If you then dig a little deeper—with the hole still the same size across the top—the sides will become steeper, and at first everything will go well. But when the sides reach a certain angle with the horizontal, called the *angle of repose,* the sand grains will begin to tumble down into the hole, making the sides a little less steep again. The sand

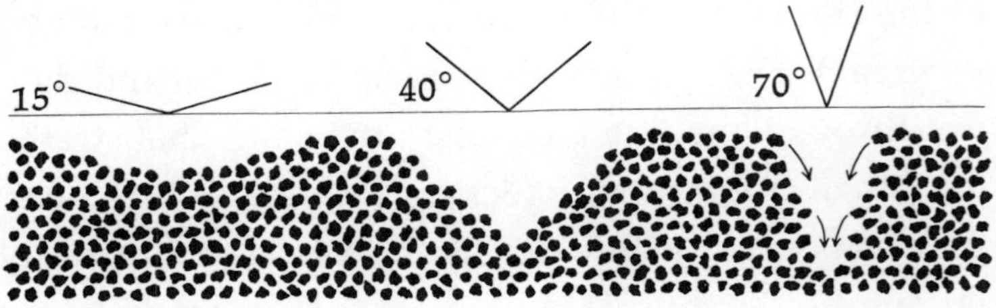

If a shallow pit is dug in dry soil (left) the sides of the pit are stable. If the sides of the pit are very steep (right), the walls will collapse as particles fall into the pit. Walls at the angle of repose (center) are stable only until they are disturbed, and then some particles will fall. The angles of the walls of the central pit are at 40° with the horizontal—the angle of repose for dry silt.

grains remaining on the wall of the pit are now precariously balanced. If you try to lay anything on them, this will disturb their balance. Some of the sand grains and your object resting on them will fall to the bottom.

The angle of repose for dry sand is about 33° with the horizontal, for dry silt it is about 40°.

The construction of an ant lion's pit is not so precise as that of a spider's web. The abdomen of an ant lion larva is pointed toward the rear, and the larva has the habit of proceeding backwards just under the surface of the sand. The streamlined abdomen acts like a plow, pushing sand over and around the larva. Some of the sand is pushed onto the larva's flat head. The larva tosses this aside by flicking its head up. At times, the larva proceeds along what appear to be aimless paths; this movement is called *doodling* (and the larva is called a *doodlebug*).

When the ant lion larva constructs a pit, it does so with movements much like doodling, but it moves in only one small area. The larva first moves in a circle, then in smaller circles inside that one—all the time flicking away the sand particles that its abdomen has plowed onto its head. Gradually a pit forms. Eventually the sides of the pit reach the angle of repose. Now any disturbance will cause some sand grains to come tumbling down. With the pit completed, the ant lion larva buries itself in the sand at the base of the pitfall and waits.

When an ant or other small insect walking over the sand reaches the edge of the pit and starts to walk down the side, a few sand grains are likely to slip out from under its legs, and a miniature landslide results. The ant, now having lost its footing,

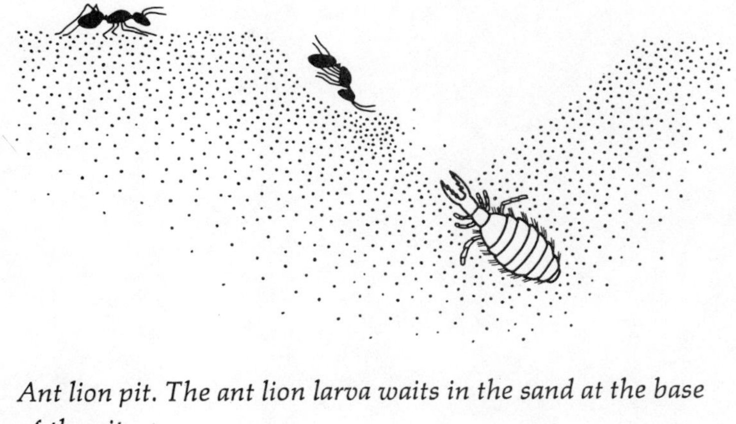

Ant lion pit. The ant lion larva waits in the sand at the base of the pit.

falls to the bottom of the pit, where the ant lion larva captures it. If the ant is faster than the ant lion and tries to run back up the sides of the pit, this will only disturb more sand grains, and the ant will tumble to the bottom again.

On the abdomen of the ant lion larva are hairs that point forward. If an insect tries to free itself from the larva's grasp, the hairs help to anchor the larva in the soil, and they also prevent a large insect from pulling the larva out of its pit.

2

Carnivorous Plants and Predacious Fungi

Carnivorous means "meat-eating." The spiders and other animals discussed in Chapter 1 are carnivorous; they eat other animals or parts of them. Carnivorous plants do not actually *eat*, but they do trap insects and other small animals, digest some parts of them, and absorb the digested material. It is the digesting and absorbing done by carnivorous plants that are similar to the digesting and absorbing done by animals—not the eating. In addition to catching small animals, the traps of carnivorous plants function a little like the stomachs and small intestines of animals.

Animals must eat food because it provides them with the nutrients that their bodies need. Among these nutrients are carbohydrates and fats (both of which are energy sources), proteins, vitamins, and minerals. As long as green plants receive enough sunshine, most of them manufacture all the carbohydrates, fats, proteins, and vitamins that they need. They also get all the minerals they need from the soil; but in acid bogs and some other swampy places, nitrogen is scarce or is difficult for plants to absorb from the soil. Without nitrogen, plants cannot manufacture proteins or some vitamins. Some bog plants have adapted to this by becoming carnivorous. Animal bodies are rich in nitrogen, and a plant that traps and digests animals gets more nitrogen than it would otherwise.

The carnivorous plants are flowering plants, and their traps are leaves or parts of leaves. The traps are either active or passive. *Active traps* have moving parts that enclose a small animal when it arrives and activates some working part of the trap. Venus's-flytraps and bladderworts have active traps. *Passive traps* have no moving parts, or if they do, the moving parts do not actually trap the victim; the victim merely blunders into a trap that has no easy exit or that has sticky surfaces that hold him. Pitcher plants, butterworts, and sundews have passive traps.

THE VENUS'S-FLYTRAP

The visible part of a Venus's-flytrap plant is a cluster of leaves each of which arises from an underground

(*a*) *Venus's-flytrap.*

stem. There is one trap at the tip of each leaf. The trap has a *midrib*, a vein that divides it into right and left sides. On the edges of the trap is a single row of long, stiff, pointed *guard hairs*. On the upper surface of the trap are six short, fine, *trigger hairs*—three on each side of the trap.

If an animal touches two of the trigger hairs within a few seconds of each other or one trigger hair twice within a few seconds, the trap will close. The opposite edges of the trap quickly come near each other, and the guard hairs on either edge alternate with each other. The trap has become a small, barred prison with the victim locked inside. At this point there is still a little space between the

(b) Open leaf. Two of the fine trigger hairs can be easily seen. The four remaining trigger hairs can be seen, but they are partly obscured by the larger guard hairs.

leaf edges, and you can look between the bars.

If you watch a closing trap but do not watch it

(c) *A leaf that has just closed. The guard hairs of the two sides interlock, but there are open spaces among them.*

very carefully, you will get the impression that the two sides of the trap move toward each other as if

(*d*) *The edges of the leaf are tightly sealed, and the guard hairs no longer interlock.*

they were hinged at the midrib. Actually, this is only a small part of the movement; most of the closing is

The closing of a Venus's-flytrap leaf is due largely to a change in the shape of the leaf. When the leaf is sealed (right) digestive fluid is secreted into the trap. These drawings correspond to parts (b), (c), and (d) of the preceding illustration.

due to a change in the shape of the two halves of the trap. In an open trap, each half is curved, with the convex side up—this is the side with the trigger hairs—and the lower side is concave. When the trigger hairs are touched, this is reversed; the upper side becomes concave, and the lower side becomes convex. This brings the edges close together, and the guard hairs mesh with each other.

If the trap is triggered by something that has no nutrient value—if you touched the trigger hairs with a pencil point, for example—the trap closes no further and within a few hours it will open again. But if it is triggered by an ant and if the ant is caught in the trap, then as the ant struggles, it touches the trigger hairs again and again. This keeps the trap closed. Gradually, the two sides of the trap move closer together until they form a tight seal all the way around the victim; you can no longer see inside. The complete sealing takes several hours.

The upper surface of the trap (which is now on

the inside) is covered with many minute *digestive glands* that produce digestive enzymes. Once a live victim is caught and sealed in the trap, the glands release a fluid containing these digestive enzymes. Because the trap is sealed, the digestive fluid does not not drip out. The enzymes kill the victim and digest some of the substances in its body. Proteins are especially rich in the nitrogen the plant needs, but proteins cannot be absorbed directly through the surface of the trap. The enzymes digest the proteins to amino acids, which also contain nitrogen. Amino acids are soluble in water, and when they dissolve in the digestive fluid, they can be absorbed by the leaf.

Digesting all the useful material from a trapped insect usually takes several days. Then the trap opens slowly. The remains of the victim—which probably are not much more than its skeleton—may stay on the open trap for a while but usually fall off eventually. The trap can now catch a new victim, if one comes along.

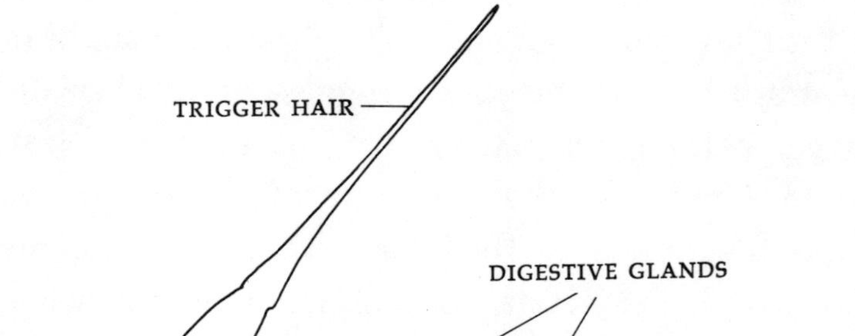

Upper surface of a Venus's-flytrap leaf.

If a trap has been sprung several times, it will close slowly or will not close at all. Old leaves open and close more slowly than young ones.

Despite its name, a Venus's-flytrap does not catch many flies. Walking and jumping animals, such as ants and spiders, are the most common victims. Agile fliers are caught only occasionally.

BLADDERWORTS

One of the most unusual traps anywhere in the plant and animal kingdoms is the trap of the bladderworts. Most of the bladderworts grow floating or submerged in water with only their flowering stalks rising above the water surface. A few grow in wet peat moss along the edges of bogs.

The submerged portion of a bladderwort consists of a stem and its branches, some of which are so fine they look like green threads. On some of these fine branches are the traps, which are believed to be modified leaves. To the naked eye the trap appears to be a small green or black bladder—black if it is filled with the dead remains of victims. Most bladderwort traps are not more than about 3 millimeters (about 1/8 inch) in diameter; some are so small they can barely be seen with the naked eye. Because of their size, the traps catch only very small animals such as water fleas, the larvae of some aquatic animals, and single-celled organisms.

At one end of the trap is an opening closed by a

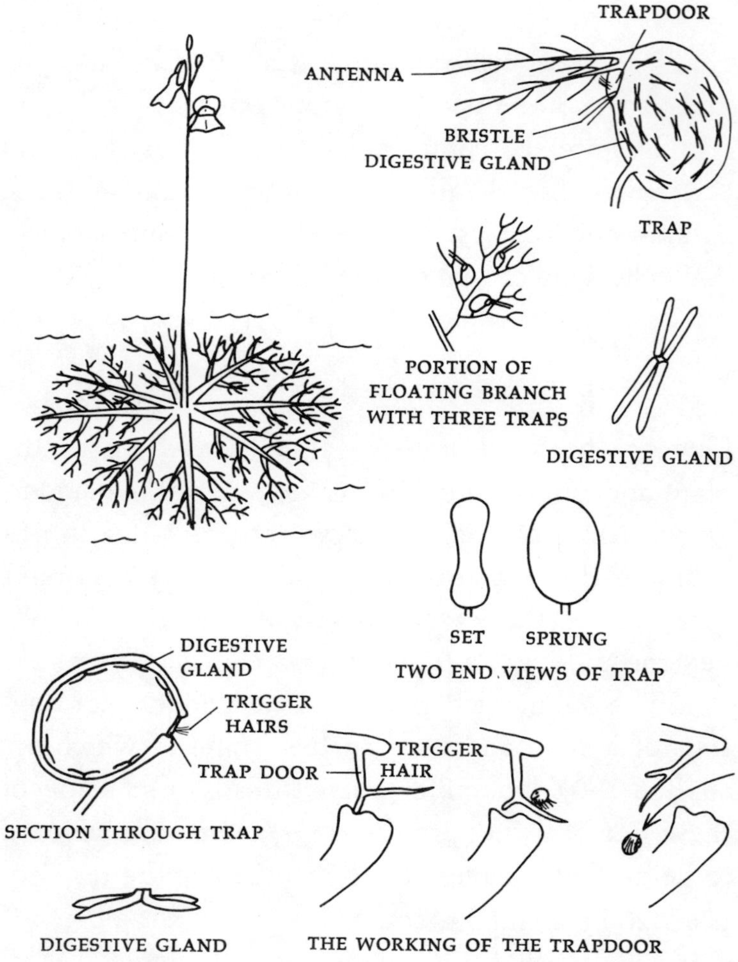

Bladderwort plant and traps.

trapdoor. The door can swing inward as if it were hinged at the top. Across the bottom of the opening is a ridge of tissue against which the trapdoor is braced. The trap is ordinarily filled with water. Its walls absorb some of this water inside the trap. This reduces the quantity of water in the trap, and the greater pressure of water on the outside causes the wall of the trap to curve inward. The water outside also presses against the trapdoor, which would swing inward, were it not braced against the tissue at its base.

Surrounding the opening of the trap and pointing forward are several hairs—two long, branched ones called *antennae* and several shorter ones called *bristles.* These hairs act as a funnel that guides small animals toward the trapdoor. In size and shape, the antennae and bristles resemble some of the filamentous algae on which microorganisms (microscopic plants and animals) often grow. Some of these microorganisms also grow on the antennae and bristles of the bladderwort trap. Here they act as "bait" that attracts water fleas and other small animals, which come to graze on the microorganisms. These animals often follow along the antennae or bristles directly to the trapdoor.

On the trapdoor itself, there are four fine *trigger hairs.* An animal that has reached the trapdoor probably will touch the trigger hairs. This bends the lower part of the trapdoor, and a small space opens between the door and the tissue below. This allows water to move into the trap. Because the pressure of water is greater outside the trap than inside, water now rushes into it, pushing the door inward and carrying the triggering animal and any others nearby along with it. The inrush of water causes the sides of the trap to swell outward. The pressure is now equal inside and outside the trap, and the door returns to its original position. All this takes only a fraction of a second.

Because the door is so constructed that it swings only inward and not outward, any animals inside are trapped. They cannot force the door open by pushing against it.

The inner surface of the trap is lined with glands that secrete digestive enzymes. While the victims are being digested, the wall of the trap absorbs more of the water within it along with whatever digested material has come from the victims. The walls once again curve inward as the pressure inside the trap becomes lower than that around it. The trap is now set again, even though the animals inside it may not be fully digested.

PITCHER PLANTS

The traps of pitcher plants are of the pitfall type. The opposite edges of a leaf are fused together most of the way up, giving the leaf a pitcher shape. The fused edges form a *wing* along one side of the pitcher. The upper part of the leaf forms a *hood,* and near it is an opening that is easy to enter but difficult to escape through. The sizes and shapes of the pitchers, wings, and hoods vary with the species of pitcher plant. Only two species are described in this book.

The most widely distributed pitcher plant of North America is the *northern pitcher plant*. Where the hood meets the rim of the pitcher, there are glands that secrete minute drops of nectar. These are too small to be easily visible, but they do attract insects. The hood is lined with stiff, downward-pointing hairs. A flying insect approaching the nectar is most likely to land among the hairs of the hood. Because the hairs point down, it is difficult for the insect to walk upward among them; it is also difficult for it to take off and fly away. About the only thing it can do is to walk into the pitcher itself.

The inner wall of the upper half of the pitcher is lined with a waxy substance so smooth it looks

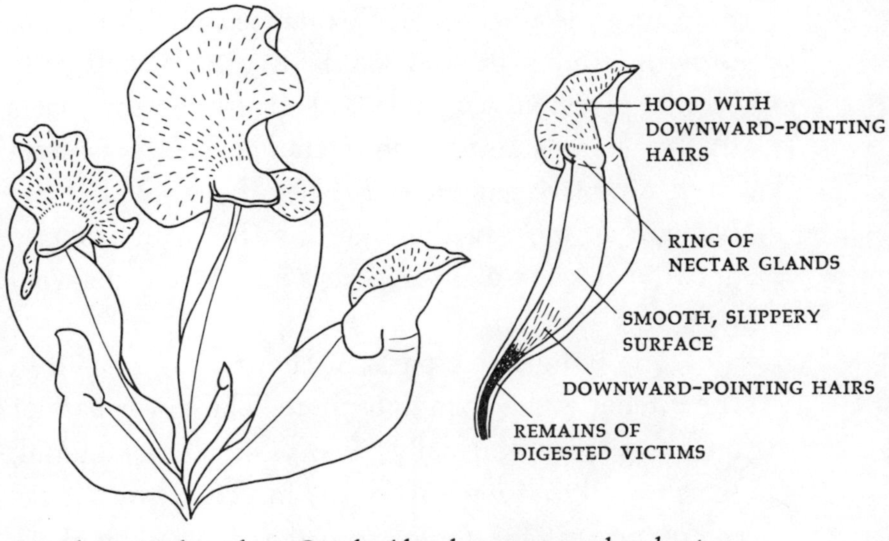

Northern pitcher plant. One leaf has been removed and cut open to show the inside of the pitcher.

like a newly polished floor; it is also as slippery. Any insect walking down from above is likely to slip and fall deeper into the pitcher. If it tries to walk back up to the opening, it just slips and falls back down again. A few flying insects are able to fly out of the opening, but because the pitcher is rather narrow, there is not enough room for any but the smallest insects to take off. Those that do attempt to fly usually hit the sides of the trap and fall down again. Insects that make repeated attempts to escape soon become exhausted.

Farther down, in the lower half of the pitcher, the wall is not waxy, but there are additional downward-pointing hairs. Here the pitcher is quite narrow, and there is almost no possibility of escape. The wall here also contains glands that secrete a fluid into the pitcher. The exhausted victim usually drowns in this fluid. Bacteria live here, and they secrete digestive enzymes that digest the soft parts of the victim's body. The pitcher itself may secrete some digestive enzymes, but it is believed that the bacteria do most of the digesting. Both the bacteria and the pitcher plant absorb some of the digested food.

The indigestible parts of the victims remain in the pitcher and accumulate there. The lower part of an old pitcher is usually filled with black material, which is all that remains of the insects. Only the remains nearest the top—those that have been there the shortest time and therefore the parts that have been subjected to digestion for the shortest period of time—can be recognized as identifiable parts of insects or other small animals. Occasionally pitchers are found that are filled to the top with dead victims.

Although the pitchers of pitcher plants would seem to be very dangerous places for small animals, there are a few that actually live inside the pitchers. There is a species of crab spider that spins its web in the upper part of the pitcher and captures animals when they fall into it. The web is a trap within a trap.

The *California pitcher plant* or *cobra lily* grows only in northern California and western Oregon. Its

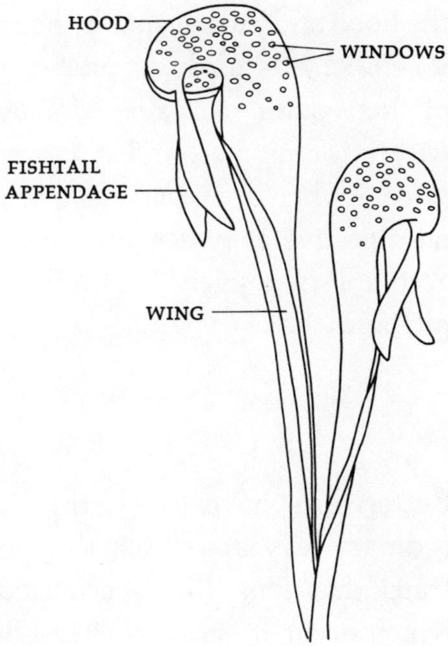

California pitcher plant.

pitchers are slender and tall. The hood bends over the top of the pitcher, and the opening to the pitcher is actually under the hood. An appendage shaped like a fishtail hangs down from the hood; this appendage and the upper part of the pitcher secrete nectar that attracts insects.

The trap is green and may be tinged with red. The upper part of the trap—especially the hood—has little patches with no color. These patches, called "windows," admit more sunlight than do the colored areas of the trap. To the insect below the opening to the hood or already inside the pitcher, these windows must look like holes through which it might fly. But the insect that flies toward one of these windows merely bumps against the inside of the hood and falls deeper into the pitcher.

Like the northern pitcher plant, the California pitcher plant contains downward-pointing hairs within the pitcher, and it also secretes fluid into the pitcher when a victim falls into it. The fluid contains no digestive enzymes, and the digestion is probably accomplished solely by the bacteria that live in the pitcher.

BUTTERWORTS

Butterworts have leaves that look very much like ordinary leaves; at least they are not modified into traps that *look* like specialized traps. The leaves, which occur in small clusters, lie flat on the ground or very close to it. Except during the flowering season, when there may be one or a few flowers at the tip of a long, upright stalk, the plants are quite inconspicuous.

There is one thing different about butterwort leaves, however. If you look very closely at the upper surface of a butterwort leaf, you will see that it is covered with what appear to be tiny dots, each barely the size of a fine pinpoint. Each of these dots

Butterwort plant.

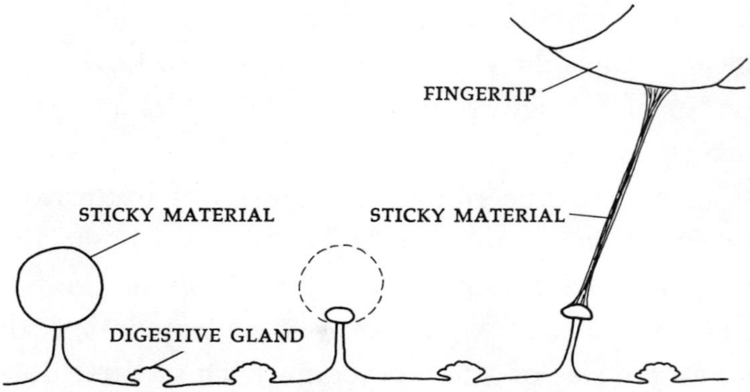

Upper surface of a butterwort leaf. The stalked gland to the left bears a large drop of slimy, slightly sticky material. The stalked gland in the center is drawn as if this material were transparent. A human finger that touches a gland can easily draw away, but a small insect would be held fast.

is a drop of slimy liquid coming from a small, stalked gland on the leaf. If you rub your fingertip over the leaf, it will feel a little like rubbing it on a piece of butter. This is why the plants are called butterworts.

If the leaf you choose is young, when you withdraw your hand, you will see a little strand of the slimy material being drawn away from the leaf. Although this material is not sticky enough to hold your finger fast to the leaf, it is sticky enough to catch small insects. Once a victim is caught, the edges of the leaf slowly curl over it, but the stickiness, not the curling, traps the victim.

Smaller than the stalked glands are stalkless glands on the leaf surface. These latter glands secrete digestive enzymes. Among the enzymes secreted by the butterwort leaf is one that causes milk to curdle. Laplanders have a food that resembles junket, which they make by mixing butterwort leaves with reindeer milk.

SUNDEWS

The several species of sundew have leaves of different shapes, but the traps all work in the same way. The *roundleaf sundew* and several other species of sundew have a small rosette of leaves that lie nearly flat on the ground, and the blade of the leaf, which

Sundew plant.

is the trap, is round or nearly round. The upper surface of the trap is covered with glands, each of which secretes a drop of fluid. The glands look like tiny dewdrops, and they glisten in the sun—which is how the plants got their common name. Most of the glands are on stalks, those near the edge of the trap having the longest stalks and those near the center having the shortest stalks or none at all.

The glands secrete a nectar that attracts insects and a sticky substance to which small insects become stuck. Once an insect has been caught, the long stalks (those near the edge of the leaf) bend slowly toward the victim. If the animal is not strong enough to get away—and larger ones do escape—the glands come to rest on it. The victim is then in a small cage composed of the leaf blade below and the long-stalked gland above and around it. The movement of the stalked glands is slow, and so, as in the case of butterworts, it is only the sticky material that does the trapping.

The glands then secrete digestive fluids, and the victim is digested over a period of several days. When digestion is complete, the stalks return to their original position.

Some species of sundew have leaf blades (traps) that are long and narrow. Most of these leaves stand upright rather than lying horizontally on the ground, but the blades are covered with glands, and they function in the same way as those of the roundleaf sundew.

PREDACIOUS FUNGI

Fungi lack chlorophyll, the green pigment that is important in photosynthesis, the process by which plants manufacture food. Lacking chlorophyll, fungi cannot make their own food. Some fungi get their food by living on or in other living things and absorbing food from them. A few fungi, called *predacious fungi,* live in the soil and trap microscopic or nearly microscopic soil animals. The word *predacious* means catching and eating animals. Predacious fungi do catch and digest small animals, but they do not *eat* them any more than carnivorous plants do. Like the carnivorous plants, predacious fungi have either passive or active traps.

The body of a predacious fungus consists of fine threads called *hyphae* (pronounced HI-fee) that branch occasionally. In a few species, the hyphae themselves are sticky and trap amebas and other small animals that happen to touch them. Other predacious fungi have special short branches on which are traps that catch nematodes. *Nematodes* are small worms, many of which live as parasites on plants and animals. Almost any soil sample contains some type of nematode. Most nematodes are barely visible to the naked eye, but they are several times thicker than the fungal hyphae that trap them.

Three common types of nematode-catching

traps on fungal hyphae are *adhesive knobs, non-constricting loops,* and *constricting loops.* Of these, only the constricting loops are active traps.

The fungi with adhesive knobs have branches that are short hyphae, each of which ends in a spherical knob covered with a sticky material. If a nematode touches one of these adhesive knobs, the nematode becomes stuck to it. Even though the nematode struggles to free itself, the sticky material is strong, and the animal is held fast. As it thrashes about, it is very likely to touch more adhesive knobs, and so it is held even more securely. Within an hour or so the nematode dies, and an *infection knob*

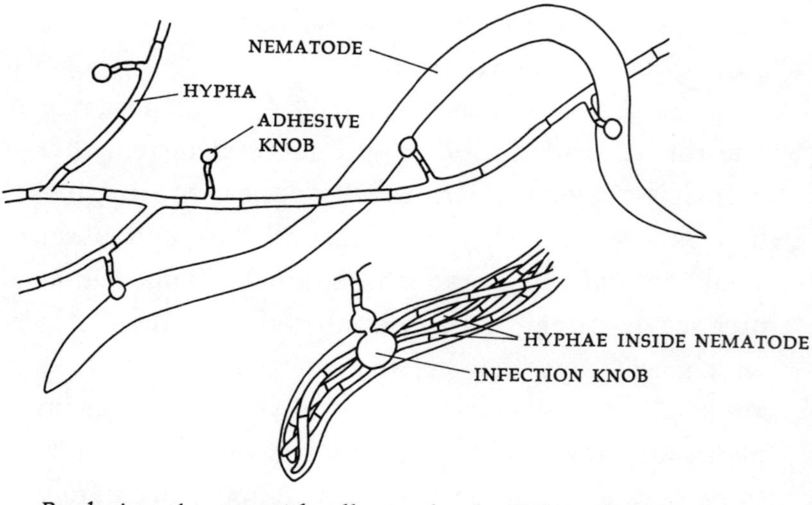

Predacious fungus with adhesive knobs. When an adhesive knob touches a nematode, an infection knob develops inside the animal. From the infection knob grow hyphae that digest the internal organs of the nematode.

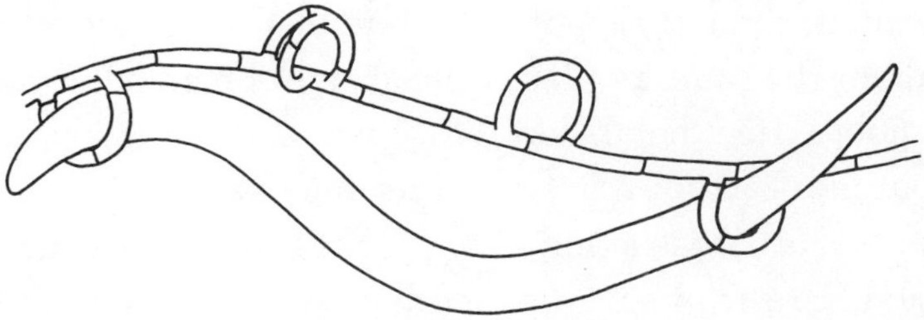

Predacious fungus with nonconstricting rings.

forms inside the animal near the point of contact with the adhesive knob. New hyphae begin to grow from the infection knob, into all parts of the dead nematode. Here they secrete enzymes that digest the soft inner parts of the animal. The hyphae then absorb the digested food from the dead body. Eventually all that is left is the hard outer integument of the nematode.

Fungi with nonconstricting rings have short hyphae, the ends of which grow around into a loop. In some of these fungi the loops are simple, each one appearing as a single ring. In others several loops may grow from an original loop forming a group of rings set at right angles to each other. The openings in nonconstricting loops are large enough to admit either end of a nematode (which tapers at both ends), but it is not large enough for the main portion of the nematode body to pass through. Therefore, once a nematode starts through a loop, it is likely to become jammed into the loop. If the nematode thrashes about, its tail end may become caught in another loop. The fate of the nematode is about the same as that of one caught on an adhesive knob. After the animal dies, hyphae grow throughout the dead body and digest its inner parts.

The constricting loops of nematode-catching fungi are formed on short hyphae. Each loop consists of three cells, and the opening in the loop is large enough for a nematode to pass through. However, nematodes move by wiggling their bodies from side to side, and they cannot pass through one of these loops without touching the inside of the loop. When this happens, the three cells of the loop swell,

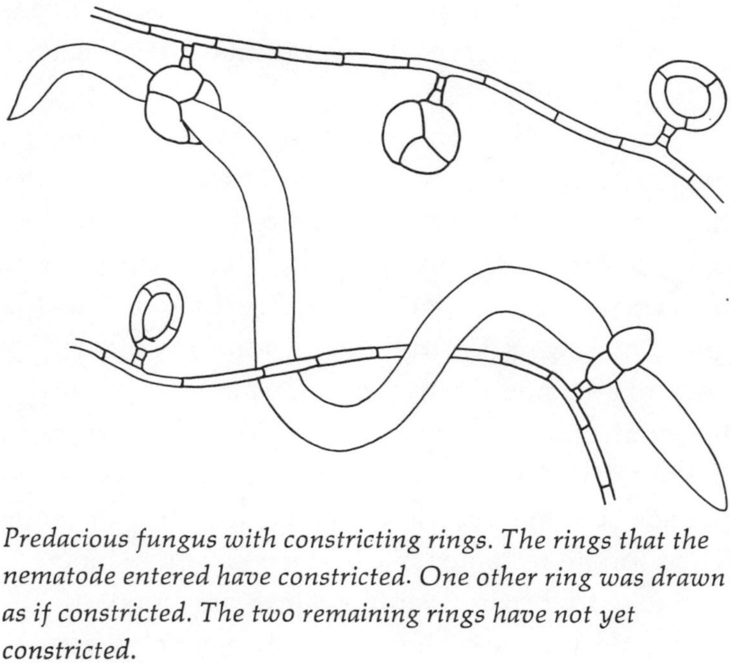

Predacious fungus with constricting rings. The rings that the nematode entered have constricted. One other ring was drawn as if constricted. The two remaining rings have not yet constricted.

and this constricts the opening in the loop. The nematode is caught as if in a tightening lasso—but it is a lasso that the animal itself entered, not one that was tossed over it. Again, the nematode is held fast, and in its struggles to escape, it may put its tail end into another constricting loop and be even more securely held. After the animal's death, hyphae grow into the dead body and digest the soft tissues.

Only the inside of a constricting loop is sensitive to touch. If a nematode brushes by the outside of a loop, the loop does not constrict. Constriction is very fast. It is usually accomplished within a tenth of a second—not enough time for the nematode to back out after becoming aware of the danger.

Just what kills the nematode trapped by predacious fungi is not known with certainty. It is believed that the fungus secretes a toxin that kills the animals. The hyphae growing into them do not kill them, for the hyphae do not enter until the animals are dead.

Although these three types of traps hold nematodes fast, nematodes do occasionally "escape" by struggling so furiously they break the hyphae. However, these "freed" nematodes still have the traps themselves stuck to them; the animals soon die—presumably by the toxin secreted by the traps—and hyphae grow into the dead animals just as they would had the animals not torn the traps from the main part of the fungus.

Whether a nematode-trapping fungus forms

traps or not depends on the presence of the nematodes themselves. If these fungi are removed from the soil and are grown in a dish in a laboratory they will not form traps unless nematodes (or a few select chemicals) are added to the dishes. Some substance secreted by the nematodes causes the fungi to begin forming traps.

Nematode-trapping fungi catch so many nematodes that some researchers feel that the traps may secrete some substance that attracts nematodes to them, but no such substance has yet been discovered. Even if the fungi do not produce an attractant, nematodes are present in the soil in such tremendous numbers that many of them would surely blunder into traps even without being attracted to them.

3

Tender Traps

Thus far we have considered only traps that kill the animals they catch or lead them toward their deaths. There are some traps, however, that hold their "victims" for a short while and then release them alive. A few flowers are such traps, and the insects caught in them are the pollinators of these flowers.

Before it can set seed, a flower must be pollinated. Pollen is produced in the *stamens* of a flower —more specifically in the *anthers,* which are sacs at the tops of the stamens. The lower portion of the stamen is the *filament,* which supports the anther. Pollen must be carried to the pistil of a flower if fruit is to be set. The *stigma,* which is the upper part of the pistil is the part that receives the pollen. The lower part of the pistil, called the *ovary,* ripens into a fruit only if pollen has reached the stigma. Between the stigma and the ovary is the *style.*

In most cases, few seeds are produced if the flower is pollinated with its own pollen. More and healthier seed usually is set if the pollen comes from flowers of another plant of the same species. This is called cross-pollination. Most flowers have some mechanism that makes cross-pollination not only possible but probable. In *trap flowers* (and in many others as well), the pistil is receptive to pollen and the stamens shed their pollen at different times. Trap flowers hold the insects captive during the time that this changeover occurs.

Only live insects can fly from flower to flower and bring about cross-pollination. If the traps killed the insects, the insects could not cross-pollinate the flowers. Not only do insects remain uninjured in most trap flowers, many trap flowers produce nectar or other food that nourishes the insects while they are there.

SWEET-SHRUB

The sweet-shrub or strawberry bush grows wild in California, but is cultivated in gardens elsewhere. It is pollinated by a beetle.

Each flower has many brownish-purple petals. Inside of the circle of petals are many stamens. And inside of the stamens with pollen are sterile stamens, called *staminodia;* these produce no pollen. At the center of the flower are several pistils. The tips of the innermost petals and of the stamens, staminodia,

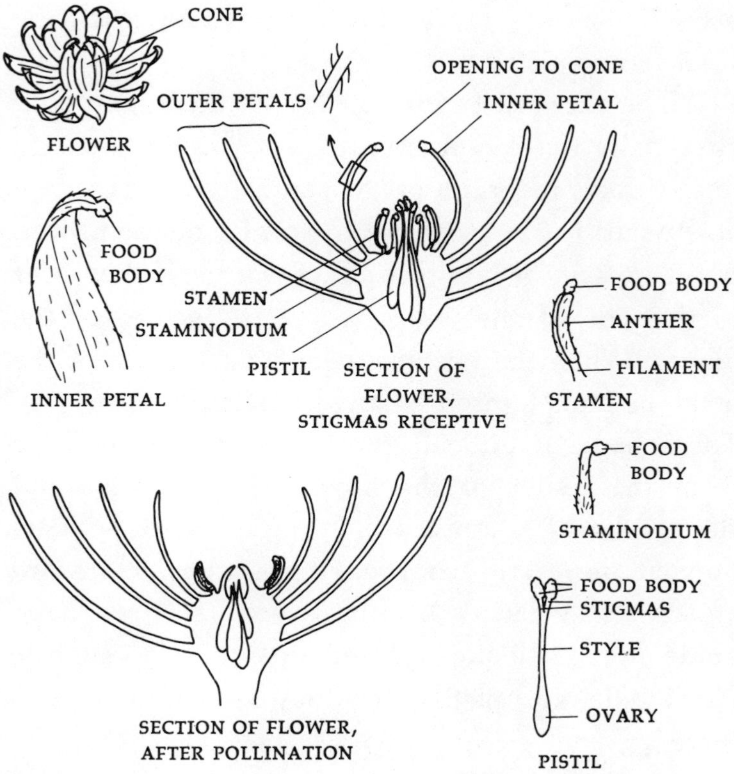

Sweet-shrub flower.

and pistils are modified into *food bodies* on which the beetles feed. In addition, the flowers produce a winelike scent that attracts the beetles.

When the flowers open, the stigmas are receptive to pollen, and the flower is so constructed that a visiting beetle is very likely to reach the stigmas first. Although the outer petals spread outward, the inner ones arch over the center of the flower and form a *cone* over it. Only a small opening among the petal tips allows the beetle to enter and reach the stamens and pistils.

The outward-facing surfaces of the inner petals have hairs that point upward. If a beetle lands on this surface, the hairs prevent him from proceeding downward (or away) from the center of the flower, but it is easy for him to walk upward and then over to the side of the petals facing the center of the flower. When the beetle reaches this surface of the petal, he finds himself among hairs that point down. These make it difficult for him to climb back up from the center of the flower. The stamens and staminodia have upward-pointing hairs; these hairs prevent the beetle from moving further downward among these organs. In other words, all these hairs guide the beetle toward the tips of the stamens, staminodia, and pistils, and they keep him trapped there.

As the trapped beetle walks around, feeding on the food bodies in his tiny prison, any pollen he has brought with him from another flower will rub off into the stigmas. When the flower first opens, the anthers are still closed; because they are not yet shedding pollen, there is little likelihood at this time that the stigmas would be pollinated by pollen from the same flower.

After the stigmas have been pollinated, the staminodia bend over them. This prevents the stigmas from receiving pollen from the anthers, which now open and shed their own pollen. Still feeding on food bodies, the beetle becomes newly dusted with pollen.

Then the remaining food bodies wither, and the beetle has nothing to eat. The flower stops producing its winelike scent, and the inner petals bend outward. No longer trapped and no longer attracted by either food or scent, the beetle is free to leave. If other sweet-shrub flowers are opening now and emitting their winelike odor, the beetle will move on to one of them. Unharmed and well fed, it will continue to visit flowers, depositing at each one some pollen from the previous flower and picking up fresh pollen.

The length of time a beetle is imprisoned in a single flower varies from about 12 to 36 hours.

BIRTHWORT

In eastern Europe there grows a vine called birthwort. Its flower has no petals, but the sepals are fused into a long tube, which is the trap. The bottom of the tube is swollen and rounded—it has a rather pot-bellied look. Here are the short stamens and one pistil with several stigmas. The stamens and pistil are fused together.

Above its lower swollen area, the tube becomes narrow, and the inside is lined with hairs. The upper part of the tube flares out into a *flag*. The flower emits an odor that attracts small gnats.

The hairs that line the tube deserve special at-

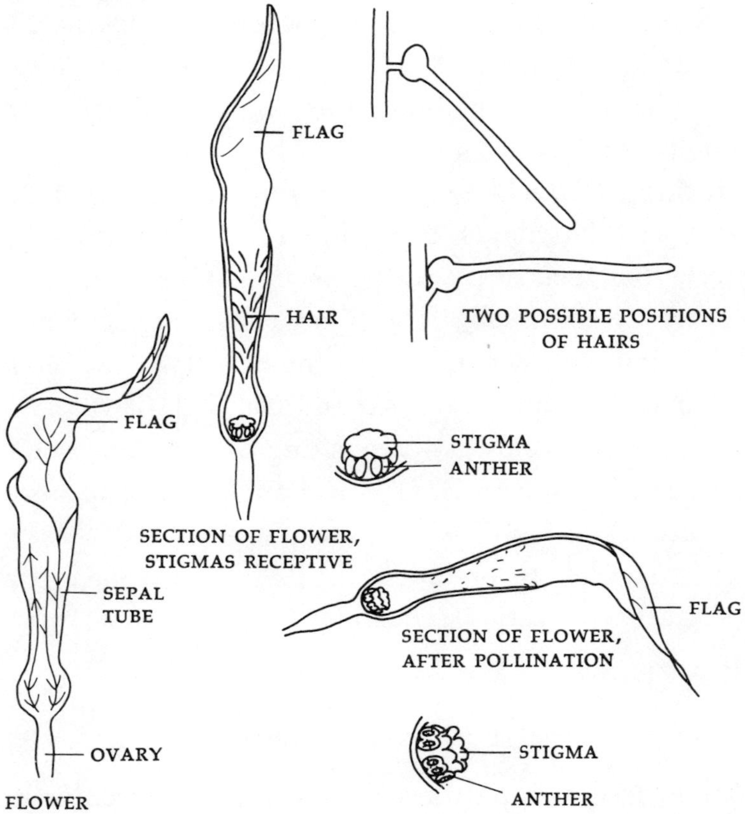

Birthwort flower.

tention. They are long, and they are attached to the wall by short, thin stalks. The stalks are flexible, and the hairs could be bent either up or down were it not for the peculiar shape of the hair just above its stalk. Here the hair bulges on its upper side, but it has no such large bulge on its lower side (though it does have a smaller one). If a gnat entering the flower pushes on a hair from above, the hair will bend downward allowing the gnat to proceed farther into the trap. If a gnat in the trap tries to escape by moving upward in the tube and pushes against a hair from below, the hair will not bend upward. The bulge will hit the wall of the tube, and the hair will stand horizontally across the tube. Since the gnat is likely to push against several of these hairs at the same time, they block his escape route.

The trap secretes nectar on which the gnats feed during the two or three days that they are trapped in the flower. As in the case of the sweet-shrub, the pistil is pollinated first by any pollen the gnats may have brought in; at this time the anthers are still closed and the stigmas spread wide over them. Later, when the stamens begin to shed their own pollen onto the gnats, the stigmas stand upright; this gives the gnats ready access to the anthers.

All this time the birthwort flower has been in an upright position. When the pollen has been shed, the flower wilts and moves into a horizontal position. The hairs wilt and shrivel. The route out of the flower is now open, and the gnats, carrying fresh pollen, can escape and visit another flower.

ARISTOLOCHIA LINDNERI

In South America there grows a relative of the European birthwort called *Aristolochia lindneri.* The flower of this plant, too, has no petals, and the sepals are fused together into a tube. The upper part of the tube flares into a flag and a long "tail"; these, together with the feceslike odor emitted by the flower, attract flies. Flies coming to the flower frequently land on the flag. The opening of the tube is in the flag, and any fly that approaches the opening too closely may fall into it.

The tube itself is bent; its shape is much like that of a children's slide in the park. It is also as smooth. Furthermore it is covered with hairs. These

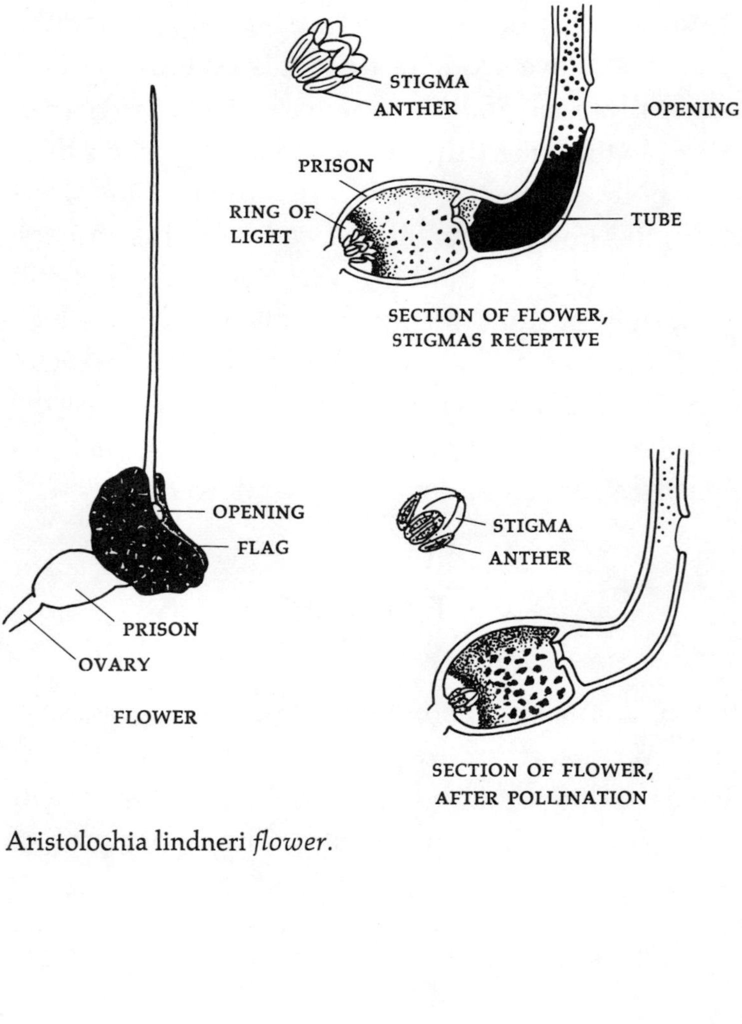

Aristolochia lindneri *flower.*

hairs are shorter than those of birthwort flowers, so they do not actually block the tube. Instead they are covered with a greasy substance that makes the walls of the tube very slippery.

The lower portion of the tube is sometimes called the *prison;* it is a chamber separated from the upper, bent portion of the tube by a cross wall. This cross wall has only one opening, and that is a small one in a short neck that points into the prison. The prison contains the pistil and the stamens of the flower.

The first day that the flower is open, the walls of the bent portion of the tube (the "slide") are dark purple, but the prison itself has some colorless areas, especially in a ring around the pistil and stamens. These areas function like the "windows" of the California pitcher plant. Because flies are attracted to light, once they have fallen into the dark tube, they voluntarily enter the prison. They are most likely to remain in the ring of light, where they walk repeatedly around the pistil and stamens. Nectar secreted by the prison provides the flies with food.

On the first day that the flower is open, the stigmas are receptive to pollen, and they are pollinated by any of the flies that have previously visited another flower. As in birthwort flowers, the stigmas bend over the anthers at this time.

The next day, the stigmas bend away from the anthers and stand erect. The anthers open and release their pollen, and the flies, still walking about, become dusted with it. After this, the flower no longer produces the scent that attracted the flies to it the day before. Gradually, the purple color of the tube fades away, making it brighter than the prison, whose walls are becoming darker as its spots grow larger. The flies are now attracted to the brighter tube, where the hairs have wilted. The opening of the prison enlarges somewhat, making it easier for the flies to leave. The flies, unharmed, will visit another flower that has just opened.

4

Femmes Fatales, *Fishing Fish, and Others that Lure*

Trapping is not the only way of catching animals. Some animals lure other animals to them by seeming to offer something that the potential prey wants. Sometimes lures are used in connection with traps, sometimes alone.

FEMMES FATALES

Femme fatale (pronounced FAHM fah-TAHL) is a French expression that has become part of the English language. It means a woman who lures men into danger. *Femme fatale* has also been applied to a female firefly that lures male fireflies to her and then eats them.

Fireflies are well known for their ability to produce a type of light called *luminescence.* Luminescence, which is also produced by a few other kinds of living things, is sometimes called *cold light.* Ordinary incandescent light bulbs become uncomfortably and even dangerously hot when they have been lit for a while, and even fluorescent lamps become noticeably warm. But luminescence is accompanied by so little heat that you do not feel it if you hold a flashing firefly in your hand. Therefore, fireflies and other luminescent organisms are not burned by their own light as you might be if you touched some light bulbs.

At the tip of a firefly's abdomen is an organ called a *lantern.* It is the lantern that produces the luminescence, which can be turned on and off. Fireflies light their tail lanterns when searching for mates. The fireflies that you may see flying about on a summer night are males. The females of most species of firefly have no wings. Even those that do have wings seldom fly. Instead, they remain on the ground, in low underbrush, or on branches of trees. Usually we do not notice the females as often as the males; but the males, which fly repeatedly back and forth about 3 or 4 feet above the ground, probably have a much better view of them than do those of us who remain on our porches.

A firefly signaling.

The males and females of each species of firefly find each other in the dark by signaling with their lanterns in a way that is different for each species. The length of a flash (from about a third of a second or less to almost a full second) and the time it takes the female to respond vary with the species. This amounts to a sort of "Morse code" by which one firefly can tell if another firefly is of its own species or not. Furthermore, the codes of males and females of the same species usually differ from each other. In this way, a firefly can also tell if another firefly is a member of the opposite sex of the same species. The signals also differ in brightness and color (yellow to greenish), but these differences do not seem to be important to the fireflies' recognition of each other. With an ordinary flashlight, you can often get either male or female fireflies to respond to you if you know the code of the opposite sex.

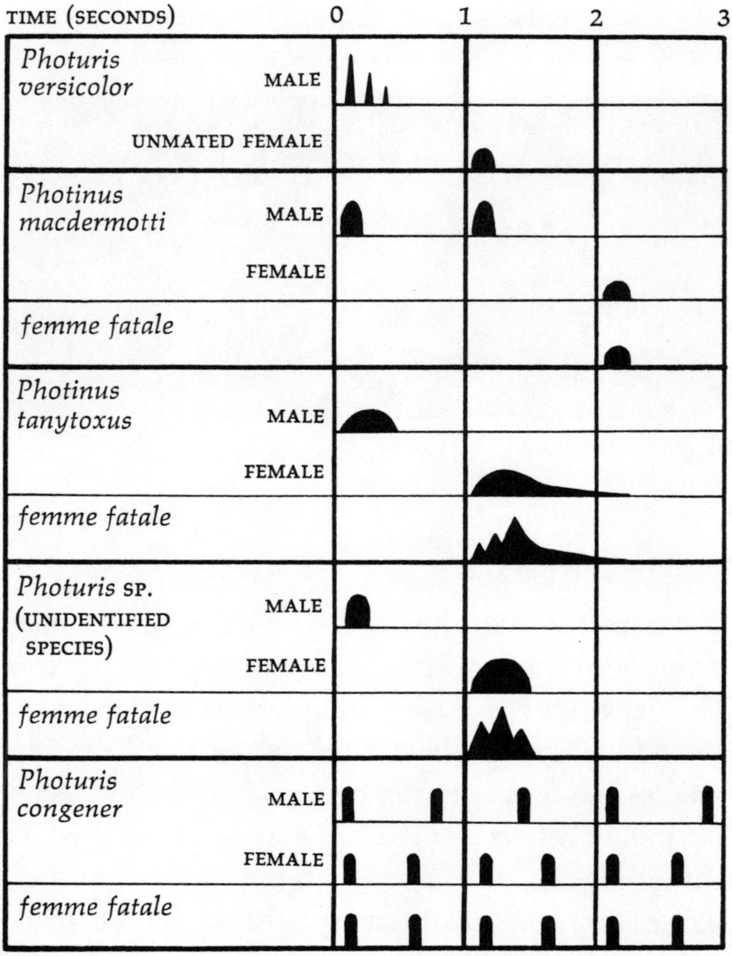

The flashing code of several species of firefly and the responses of femmes fatales, *which are the mated females of* Photuris versicolor. (*Redrawn from Lloyd, James E.* Science *187: 452–453, 7 February 1975. Copyright © 1975 by the American Association for the Advancement of Science.*)

In most firefly species, the luminescence in the adults has no function but to bring males and females together. But the females of *Photuris versicolor,* after they have mated—and only after they

A femme fatale *(left) beginning to eat the male firefly she has lured to her. (Photograph by James E. Lloyd, University of Florida,* Science, *Cover, Vol. 187, 7 February 1975. Copyright © 1975 by the American Association for the Advancement of Science.)*

have mated—begin to change their pattern of signaling. Now when they see a male of a different species signaling, they respond to him by mimicking the code of the female of his species. These mated females who mimic other females are the *femmes fatales.*

For they are the ones that lure to their deaths the males of other species.

A deceived male approaches a *femme fatale* as he would a female of his own species. As soon as he comes close, the *femme fatale* pounces on him and eats him. The deception is not always perfect, and it does not work every time. A *femme fatale* may lure to her only one out of ten males to whose signals she responds, but that may be enough to satisfy her hunger.

Most *femmes fatales* can change their flash pattern from that of one species to another quickly. Sometimes they do this immediately; sometimes they do not answer until they have seen the new male signaling several times—but usually within less than a minute. Occasionally a *femme fatale* will not respond to the signals of one particular species but will respond to others. *Femmes fatales* do not respond to the signals of the males of their own species, who are safe from them. (They did respond before they mated, but then they were not *femmes fatales.*)

Nearly every child who lives where fireflies are abundant has captured some of them and put them into a glass jar. Usually all the fireflies survive for at least a day or two. Occasionally, however, the morning after a collection was made, one firefly is alive, and all the others are dead, empty, broken shells. In such a case, the collector has unknowingly captured a *femme fatale.*

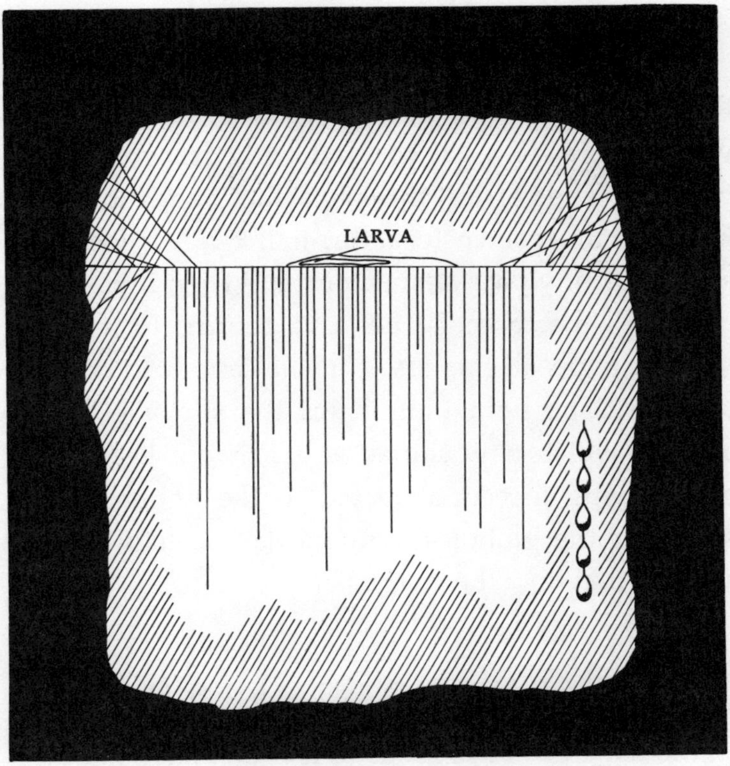

Larva of the fungus gnat of New Zealand and its trap. A small portion of one of its threads with sticky globules is enlarged to the right.

THE FUNGUS GNATS OF NEW ZEALAND

In certain caves in New Zealand there live fungus gnats, the larvae of which spin webs that are a little like those of the spider *Steatoda borealis* and also have tail lanterns that are similar to those of fireflies.

Near the ceiling of the cave, each larva spins one horizontal thread, which is anchored to the walls by a few other threads. On the horizontal thread the larva constructs a long, narrow tube that is its home.

From the horizontal thread the larva suspends many vertical threads, each covered with drops of sticky globules. The vertical threads, which range in length from about an inch to about half a yard, look like strings of tiny pear-shaped pearls. When it has completed all the threads, the larva returns to its tube.

At night all the larvae in the cave turn on their lanterns. The light illuminates the sticky globules, each of which now shines as if it had a light of its own. Flying insects attracted to the lights become stuck on the globules, and the larvae haul up the threads that hold the victims.

FLASHLIGHT FISHES

In the sea are several kinds of fish that have luminescent organs. The *flashlight fishes,* which live in shallow tropical waters, do not actually produce the light themselves. Below each eye, the fish has a pocket called a *light organ,* in which live luminescent bacteria. The bacteria glow continuously; their light is never turned off.

In one flashlight fish (*Photoblepharon*), there is below each light organ a flap of black tissue that resembles a lower eyelid. Like a lower eyelid, it can be raised, and this covers the glowing bacteria. Raising this false eyelid is called *blinking.* When the flashlight fish blinks, the light goes out—actually it only appears that the light has gone out, for the light is merely hidden behind the black tissue.

Flashlight fish (Photoblepharon) *blinking. The light organ is shown in white, and the flap of tissue that covers it is black.*

In another flashlight (*Anamalops*), one side of the light organ is transparent, and the other is covered by a black tissue. This fish controls its light by flipping its light organ over so that either the transparent or the black side is exposed.

In the darkness of night, the flashlight fishes lure prey to them by turning their lights on. It is known that the light has several additional functions, which include illuminating the fishes' own way, frightening predators away, attracting mates, and otherwise communicating with members of the same species.

ANGLER FISHES

Deep-sea fishes live in constant darkness, for sunlight does not penetrate below about 600 feet in the clearest of waters and much less than that in murky waters. Not only is it dark deep in the sea, but food is often relatively scarce there.

Several deep-sea fish, known as *angler fish*,

lure prey with luminescent bait. Angler fish have a modified dorsal fin; this is the large fin on the upper, or dorsal, surface of the animal. In the angler fishes, a part of the dorsal fin is long and slender and looks a little like a flexible fishing rod—which is just the way it is used. At or near the tip of the fin is an enlarged bit of tissue that glows; this is the "bait." The angler fish can move this dorsal fin forward, dangling the luminescent lure before it as it swims. When another fish, attracted by the light, approaches, the angler fish brings the lure closer to its own mouth until the prey is drawn in closely enough to be snapped up. In one unusual angler fish, the modified dorsal fin is about three times the length of the fish's body.

There are several angler fishes that live in shal-

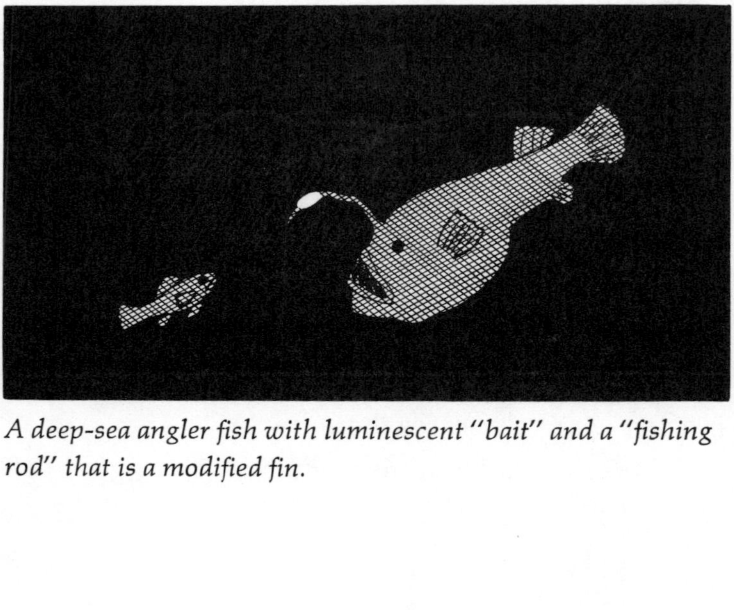

A deep-sea angler fish with luminescent "bait" and a "fishing rod" that is a modified fin.

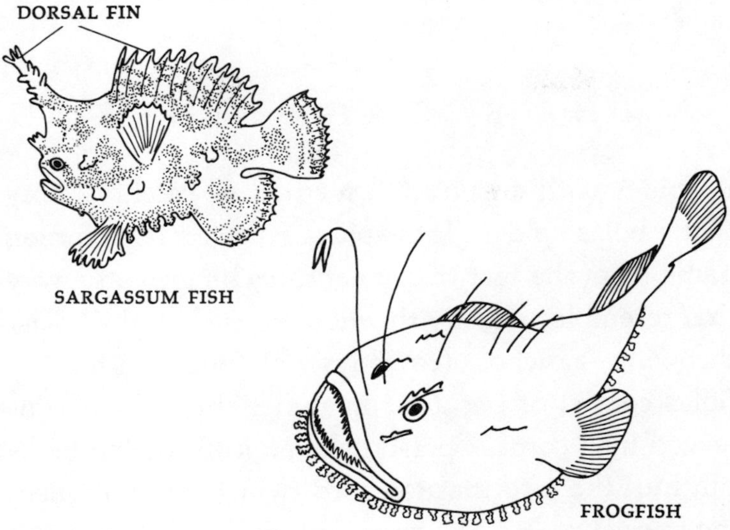

Two angler fishes of surface waters. The irregular pattern and the small finlike projections of the sargassum fish enable it to hide among floating seaweeds. Its "fishing rod" is part of its dorsal fin. The frogfish rests on the bottom of shallow waters and lures passing fish. Its dorsal fin consists only of several spines, the most forward of which carries the "bait."

lower waters and hunt by day. These, too, have dorsal fins modified into fishing rods, but the bait does not glow. These angler fish move the bait much as a human fisherman would move his rod and line baited with a minnow. In fact, the tissue at the tip of one angler fish's fin looks exactly like a miniature fish.

Many angler fishes have mouths that open upwards—just below the route that another fish might take when approaching the bait.

THE BOLAS SPIDER

A spider with an unusual method of catching prey is the bolas spider. This spider received its common name from the fact that it captures its prey in a way that resembles slightly the method used by the bolas-throwing gauchos (cowboys) of South America. Bolas consist of two (or sometimes three) balls connected by a cord. A person skilled in throwing bolas can hurl them so that the cord twines around a fleeing animal's legs. The animal stumbles and falls and can be easily caught.

The bolas spider is a member of the spider group called orb weavers, but she spins no orb web. Instead, every evening about sundown she spins a single horizontal line called a *trapeze*. The trapeze is anchored at both ends—usually to the same twig. The bolas spider hangs from her trapeze by two or three legs on one side of her body. Then she produces another line with one large drop of viscid silk at the end. This is her "bola." She holds one end—the one without the sticky ball—in her palps (the small appendages near her mouth) and holds the middle of the line with a front leg. The end of the line with the ball of viscid silk hangs free. When a moth flies near, the bolas spider swings her sticky ball in the direction of the moth. As the ball swings forward, it elongates; this probably increases the

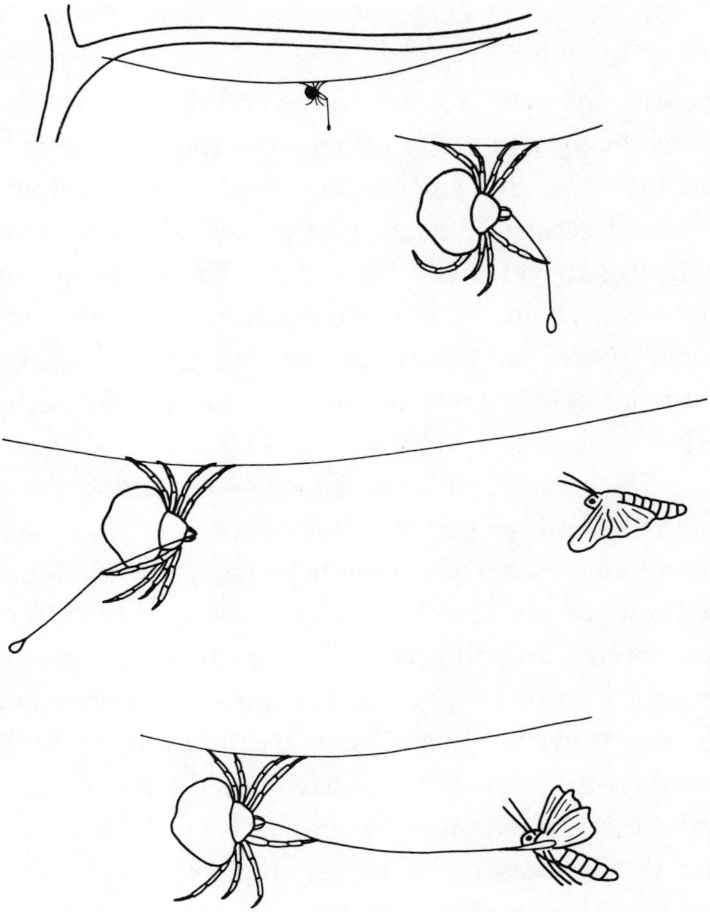

Bolas spider awaiting prey (above), taking aim (center), and striking prey (below).

chances of at least some part of it hitting its target. If the ball hits the moth, the spider hauls it in, paralyzes it with a bite, and wraps it in swathing silk.

You can see that the bolas spider does not use her bola in exactly the same way that the gauchos use theirs. She does not twist a cord around her prey;

instead, the prey is glued to the end of her line. She hauls the prey in to her rather than going to where it has fallen, as the gaucho does. And there is yet another difference. The gaucho pursues an animal that is trying to get away from him. The bolas spider lures her victim to her. Researchers working with bolas spiders are almost certain that when hunting, the bolas spider emits an odor that lures male moths to her.

The females of some animals—including these moths—emit a scented substance called a *sex pheromone*. A sex pheromone attracts an individual of the opposite sex. The sex pheromones of different species are different. This enables animals to recognize a prospective mate of the same species and usually prevents them from trying to mate with members of other species. Often it is the female that emits a sex pheromone when she is ready to mate, and this attracts males to her. If there is a breeze, the sex pheromone is blown downwind. When a male moth detects the sex pheromone, he follows it upwind, and this brings him to the female moth.

At night, when the bolas spider is hunting, she apparently emits a substance that mimics the sex pheromone of a female moth. An investigator who watched bolas spiders for several nights learned the following facts: All the moths caught by the spider were male moths. All the moths caught were of two species. (This may be because the sex pheromones of these species are similar or because the bolas spiders can produce two different substances that mimic two different moth sex pheromones.) All the male moths approached the spider from downwind, which is exactly what they would do if they were following the scents of their females. If the investigator placed a large piece of cardboard a short distance downwind of the spider to change the direction of the breeze, the male moths also changed their direction of approach and followed the breeze upwind around the cardboard. Furthermore, if the spider's aim with her ball was poor and she missed a moth, he did not flee, but continued to hover around her as if he were around a female moth. Thus the spider had a second chance to hit her target.

By day, when she is not hunting, the bolas spider holds no line or ball, and apparently she emits no imitation sex pheromone, for while she rests, no moths are attracted to her. But each evening she spins a new trapeze line and begins hunting again.

THE CRAB SPIDER OF JAVA

In Java there lives a crab spider that spins a web that is not a trap. Rather, its web is part of a lure that attracts the butterflies on which the crab spider feeds. These butterflies need salts in their diet, and they get the salts by sucking up the liquid that oozes from bird droppings.

The crab spider of Java is white with some

black markings. As long as the spider remains still on a leaf or branch it looks like a bird dropping. But before settling down, the spider spins a small web around itself. The texture of this web is such that it resembles the moisture that seeps from a real bird dropping. Thus disguised, the spider waits until a butterfly is attracted by the sight. When the butterfly approaches closely, the spider can easily grasp it.

The imitation is so good that the man who first discovered this spider and its web later held a leaf in his hand for several minutes before he realized that the "bird dropping" on the leaf was actually another spider and its web—and the only reason he became aware of it then was that the spider bit him.

FLOWERS THAT ARE SOLAR FURNACES

Animals are not the only lurers. As we saw in Chapters 2 and 3 the traps of carnivorous plants and some flowers attract animals to them. Nearly all animal-pollinated flowers have something that lures animals to them, even if the flowers are not trap flowers. In most cases the lure is something to eat—either nectar or pollen—and the flower may have a bright color or emit a particular odor that the animals learn to associate with that food. In a few cases the lure is something else. We will discuss in this section just one unusual lure, a lure that is characteristic of some flowers of cold regions—heat.

In arctic regions summer days can be cool, especially in areas near large bodies of water. Birds and mammals—the so-called warm-blooded animals—have built-in mechanisms that maintain their body temperatures, and so they are not particularly affected by cool weather. All other animals—sometimes called cold-blooded animals—have body temperatures that are very close to that of their environment, especially when the animals are at rest. Since small animals lose heat faster than large animals, very small insects are especially likely to lose heat rapidly in cool weather.

Intense physical activity warms the body, but the animal must expend a great deal of energy to produce this heat. Another way of keeping warm is to bask in a warm, sunny spot. At least two arctic flowers—*mountain avens* and *arctic poppy*—have just such warm places, and mosquitoes are attracted to these flowers even though the flowers do not contain nectar. Some mosquitoes have been timed as staying up to 13 minutes in one of these flowers; this is a long time for an insect to stay in a flower unless it is getting something from it.

The flowers of mountain avens and arctic poppy function like miniature solar furnaces. A solar furnace contains no fire. It is merely a bowl that faces the sun. As light rays from the sun strike the inside of the bowl, they are reflected toward one central place within the bowl. Anything in that spot becomes warmed because so much more light strikes

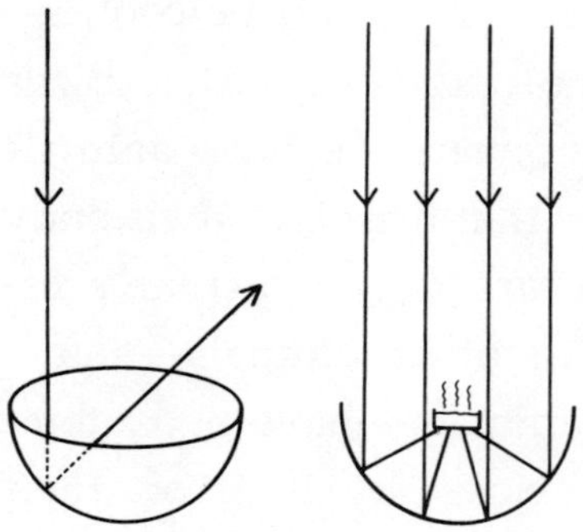

A solar furnace is bowl-shaped. Light rays hitting the inner surface of the bowl are reflected toward a central area. If the solar furnace is large enough and the light is intense enough, the heat produced in this area will boil water placed there.

it than would strike it if it were merely in sunshine but not in the solar furnace. The amount of heat produced depends in part on the size of the solar furnace. One a few feet across can focus enough light rays to boil water or cook a dinner, and a very large one can produce much higher temperatures.

The flowers of mountain avens and arctic poppy are bowl-shaped, with the petals forming the bowl. In the center of the flowers are the stamens and the pistils. The sunlight that strikes the petals is focused toward the tips of the pistils and stamens. On sunny days the temperature may be as much as 6°F higher in a mountain avens flower than in the surrounding air, and it may be up to 10°F higher in an arctic poppy. A mosquito landing in the flower basks in its warmth for a while. As it does, some of the pollen from the anthers may become attached to the mosquito's body, and pollen that it picked up on an earlier visit to another flower may be rubbed off onto the pistil.

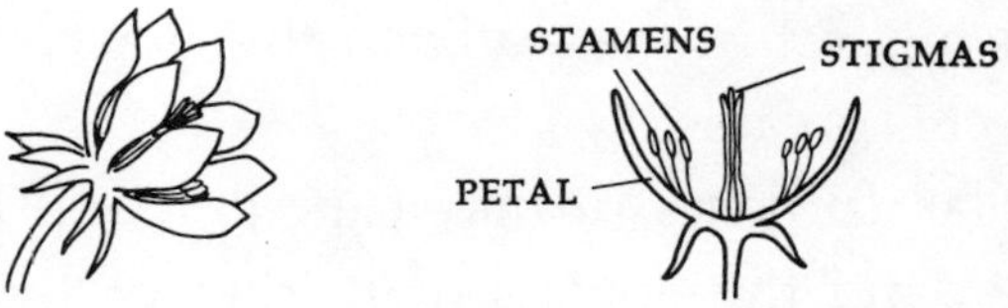

Mountain avens flower (left) and a section of the flower (right). The petals form a bowl. Sunlight striking the petals is reflected towards the stigmas. Because the flower is not a perfectly shaped bowl, some light is concentrated a little farther away from the stigmas; this area includes the anthers.

Flying insects do not fly when their bodies become cold. If they do not fly from flower to flower they do not pollinate the flowers. On cool days the warmth that the insects receive in the flowers may be just enough to make the difference between flying and not flying. On other days it may determine whether the insects are only slightly active or are very active. The more flowers the insects visit, the more flowers they will pollinate. The heat that lures the insects to the flowers thus ensures that they can be lured to even more flowers.

To function, solar furnaces must face the sun. In the high arctic summer the sun is above the horizon for 24 hours a day, and it makes a complete circle around the sky—from north to east to south to west and back to north again—in those 24 hours. The flowers of arctic poppy face the sun all day, following it as it moves around the sky. Those of mountain avens face the sun primarily in late morning and early afternoon; at this time of day the sun is at its highest in the sky, and this is when its heating effect is greatest.

Solar furnaces do not function well in cloudy weather, and the flowers of neither of these plants follow the sun when the sky is overcast.

FLOWERS THAT KILL

Flowers pollinated by animals usually do not harm these animals. It is true that an occasional trap flower may fail to open or some of its visitors may not find their way out no matter how clear the path to freedom. Some flowers are so sturdy that if an insect's leg or proboscis becomes caught on a flower part, it may not be able to free itself and will die there. A small insect may drown in a large pool of nectar or become mired in unusually thick nectar. But these are accidental happenings. In the general course of events, sooner or later, the pollinators are free to leave flowers. If they could not leave, they would not do much cross-pollinating.

In spite of this, there are some flowers that kill. It is not the pollinators themselves that die, however. Rather, it is their offspring. We are concerned here with one more unusual lure by which a few flowers attract pollinators: a seemingly good place to lay their eggs.

One of the several reasons we like to dispose of garbage quickly is that it begins to smell bad as it grows old. Another reason is that flies come to decaying food, and it is the odor that attracts them. Some flies feed directly on the garbage, but others, called *carrion flies,* lay their eggs in rotting meat. When the maggots (larvae of flies) hatch from the eggs, they require food, and they feed on the meat. Adult carrion flies are instinctively attracted to the odor of rotting meat. Ordinarily this ensures that their offspring will have plenty of food.

In the *carrion-flower* of Africa, the lure is a

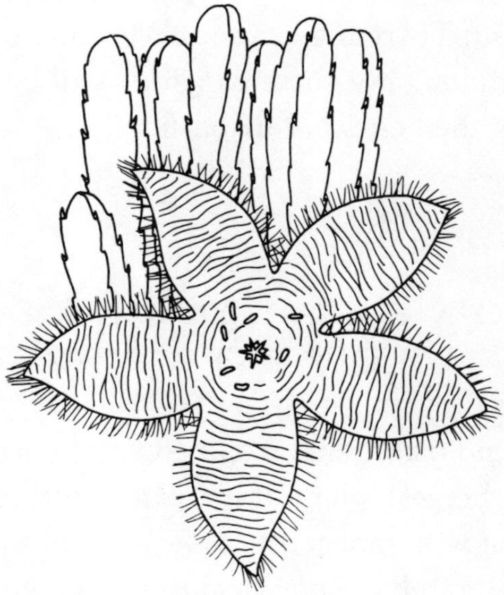

Carrion-flower with maggots.

strong odor resembling that of rotting meat; the reddish-brown flowers are also the color of meat. Adult flies ready to lay eggs react to the flowers as they would to rotting meat—they lay their eggs in them. The adult fly is free to fly to another flower and lay more eggs there. In the course of visiting several flowers, the fly transfers pollen from one to another.

This is a rather one-sided arrangement. In the case of almost all other pollinations carried on by animals, the animals receive food or another benefit. But in this case, only the plant benefits. The fly's offspring are doomed, for there is no meat in the flower, and the larvae starve to death.

The flies are not in danger of becoming extinct because of this, for many adult flies do lay their eggs in real meat. There the larvae survive and eventually become adults, only some of which will be deceived into laying their eggs in carrion-flowers.

ONE LAST WORD ABOUT TRAPS

There are no plants that trap people or even dogs or cats. The biggest plant traps catch nothing larger than perhaps a mouse or a very small frog—and these only rarely. Any vicious plants you see in jungle movies are man-made, not man-eating.

GLOSSARY

ACTIVE TRAP: a trap with moving parts that encircle or otherwise catch an animal

ADHESIVE KNOB: the trap of one type of predatory fungus; it is merely a small knob covered with sticky material

ANGLE OF REPOSE: the steepest angle at which soil particles will remain stable and not slide downward under the pull of gravity; the angle of repose is different for different types of soil

ANGLER FISH: any of several fishes in which the dorsal fin or part of it functions like a fisherman's pole, line, and bait

ANTENNAE: one of the hairs that guide prey to the bladderwort trap

ANTHER: the upper part of a stamen that holds the pollen

BLINKING: in flashlight fishes, the bringing of a flap of black tissue resembling a lower eyelid over the light organ or the rotating of the light organ so as to prevent the light from being seen by other organisms

BOLA: a cord connecting two or three balls traditionally used by gauchos to capture animals; the line and ball of viscid silk with which the bolas spider captures moths

BRIDGE LINE: the first line laid down in the spinning of an orb web

BRISTLE: one of the hairs that guide prey to the bladderwort trap

CARNIVOROUS: eating meat; most of the animals mentioned in this book—those that trap other animals and eat them—are carnivorous; carnivorous plants do not actually eat meat, but they do digest the tissues of the animals they catch

CARNIVOROUS PLANTS: green plants that trap small animals, digest them, and absorb the digested material

CATCHING THREAD: the viscid, spiral thread of an orb web; it is composed of elastic silk covered with drops of viscid silk

CEPHALOTHORAX: the fused head and thorax of a spider (in insects the head and thorax are not fused)

COBWEB: the irregular, three-dimensional web of common house spiders as well as some others; it is composed of only dry silk

CONE: in sweet-shrub, the covering formed by the bending of the inner petals over the stamens and pistils in the center of the flower

CONSTRICTING LOOP: the trap of one type of predacious fungus; the loop initially has enough room for a nematode to enter, but the loop constricts around the nematode as soon as it touches the inside of the loop

CROSS-POLLINATION: the pollination of a flower with the pollen produced in the flower of another plant of the same species

DOODLEBUG: the larva of the ant lion

DOODLING: the seemingly aimless wandering of an ant lion larva (doodlebug) just below the surface of dry sand or dry, fine soil

DORSAL FIN: fin on the back, or dorsal side, of a fish

DRAGLINE: a dry silken line laid down by spiders as they move about; it is attached at intervals and serves as a lifeline if a spider falls from a height; it is also used in some web construction

DRAGLINE SILK: a dry silk used in draglines and some threads in a spider web

ELASTIC SILK: silk that stretches; together with viscid silk it forms the viscid spiral (catching thread) of an orb web

femme fatale: a female firefly that lures to her males of other species and eats them; the plural of this expression is formed as it is in French: *femmes fatales*

FILAMENT: the stalk of a stamen

FOOD BODY: a small mass of food-rich tissue on some parts of the sweet-shrub flower; beetles feed on the food bodies while imprisoned in the flower

FOUNDATION LINE: one of the lines, other than the bridge line, that frame an orb web

FREE ZONE: zone between the notched zone and the spiral thread in an orb web; here there are no threads connecting the radii

GUARD HAIRS: the long, stiff hairs along the edges of a Venus's-flytrap leaf; the guard hairs of the opposite edges mesh when the trap closes

HACKLED BAND: a flat silken thread used in some spider webs

HOOD: the upper part of the pitcher of pitcher plants

HUB: the center of a spider web; the radii radiate from the hub

HYPHA: one of the threads that form the body of many fungi

INFECTION KNOB: a knob of fungal tissue that forms inside a nematode newly killed by a predacious fungus; from the infection knob grow fungal hyphae that digest the tissues of the dead nematode

LANTERN: the light organ of several luminescent organisms, especially fireflies and the larva of the New Zealand fungus gnat

LIGHT ORGAN: a luminescent organ in living things that produce light

LUMINESCENCE: a type of light accompanied by very little heat

MIDRIB: the main vein that runs down the center of many leaves

NEMATODE: a roundworm; most roundworms that live in the soil are much smaller than earthworms, some being almost microscopic

NONCONSTRICTING LOOP: the trap of one type of predatory fungus; a nematode trying to pass through the loop becomes wedged in it

NOTCHED ZONE: area immediately around the hub of an orb web; the radii are connected by threads in a somewhat irregular arrangement

ORB WEAVER: any of a group of spiders that spin orb webs; a few spin modified webs that do not resemble orb webs (the bolas spider, for instance)

ORB WEB: a flat spider web with the viscid (spiral) thread occupying a round area

OVARY: lower part of a pistil; the ovary contains one or more ovules that may ripen into seeds after pollination

PALP: an appendage with the mouth of some animals; in spiders the palps look almost like a pair of short legs and they may be used in holding objects

PASSIVE TRAPS: a trap that catches an animal without the use of moving parts; passive traps are sticky traps or pitfalls

PERMANENT SPIRAL: the viscid spiral or catching thread of an orb web; the permanent spiral is not actually permanent, for it is frequently replaced

PENDULUM TURNS: turns made by a spider when laying down the catching thread of an orb web; instead of being a spiral, the thread alternates between a clockwise and counterclockwise course; depending on the species of spider, all or only part of the catching thread follows pendulum turns

PHEROMONE: a substance that is produced by an individual and that causes a particular response in another individual of the same species

PISTIL: female part of a flower; it usually consists of an ovary, one or more styles, and one or more stigmas

PITFALL: a trap into which it is easy to fall but from which it is difficult or impossible to escape

PREDACIOUS: catching and eating animals; predacious fungi do not actually eat animals, but they do digest their tissues

PREDACIOUS FUNGUS: a fungus that traps very small animals, digests them, and absorbs the digested material

PRISON: a portion of the *Aristolochia lindneri* flower in which insects are imprisoned during the time they pollinate the stigmas and receive pollen from the anthers

RADIUS: one of the dry lines radiating from the hub of an orb web; the radii support the viscid spiral (catching thread)

RETREAT: a hidden area where a spider awaits the entangling of prey in her net; the retreat may be merely a spot under a nearby leaf or it may be constructed from silk by the spider herself

SEX PHEROMONE: a substance that is emitted by individuals of one sex and that attracts to them individuals of the opposite sex of the same species

SIGNAL LINE: a silken line that runs from a spider web to the spider's retreat; the struggling of newly trapped prey in the web causes the signal thread to vibrate and alerts the spider

SOLAR FURNACE: a bowl-shaped structure that produces heat by reflecting light (as a mirror does) toward one central area where the item to be heated is placed

SPINNERET: organ through which the spider forms silken threads from liquid silk produced in her silk glands; the spinnerets are near the tip of the abdomen

STABILIMENTUM: a broad, zigzag band on some orb webs; it is thought to camouflage the spider when she is on her web

STAMEN: the male part of a flower; it consists of a filament and an anther in which pollen is produced

STAMINODIUM: a sterile stamen

STIGMA: the portion of a pistil that is receptive to pollen

STYLE: a stalk extending above the ovary of a pistil; a style bears one or more stigmas; some pistils have very short styles or even none

SWATHING SILK: silk used by some spiders to wrap (swathe) their prey

TEMPORARY SPIRAL: a spiral of dry silk laid down in an orb web before the viscid spiral (catching thread) is formed; the spider destroys the temporary spiral when the viscid spiral is laid down

TRAP FLOWER: a flower that traps insects while cross-pollination occurs

TRAPEZE: the single silken line from which a bolas spider hunts at night

TRAPPING THREAD: catching thread

TRIANGLE WEB: the triangularly shaped web of the triangle spider; it resembles three adjacent sectors of an orb web

TRIGGER HAIRS: fine hairs, which when touched cause a trap to operate; trigger hairs are present in the traps of Venus's-flytrap and bladderworts

TUBE: the tubular portion of some trap flowers; the tube is a passageway that allows easy entrance to insects but prevents escape until cross-pollination has been accomplished

VISCID SILK: sticky silk, which is a part of many spider webs; victims become stuck on drops of viscid silk

VISCID SPIRAL: thread covered with viscid drops that trap victims; it is the catching thread or permanent spiral of orb webs

WING: an extension formed by the fusion of the opposite edges of the leaf that is the pitcher of a pitcher plant; the wing probably leads crawling insects from the ground up to the opening of the pitcher

Index